TEACHING MATHEMATICS IN THE BLOCK

Susan Nicodemus Gilkey

and

Carla Herndon Hunt

Routledge
Taylor & Francis Group
New York London

First published 1998 by Eye On Education

2 Park Square, Milton Park, Abingdon, Oxon OX14 4RN
711 Third Avenue, New York, NY 10017, USA

Routledge is an imprint of the Taylor & Francis Group, an informa business

First issued in hardback 2017

Library of Congress Cataloging-in-Publication Data

Gilkey, Susan (Susan N.), 1950-
Teaching mathematics in the block / by Susan Gilkey and Carla Hunt.
p. cm.
Includes bibliographical references
ISBN 1-883001-51-X
1. Mathematics—Study and teaching. 2. Curriculum planning. 3. Schedules, School. I. Hunt, Carla (Carla H.), 1946-
II. Title.
QA11.G468 1998 97-46548
510'.71—dc21 CIP
ISBN: 978-1-883001-51-3 (pbk)
ISBN: 978-1-138441-62-0 (hbk)

Other Books on Block Scheduling

Teaching in the Block
Stategies for Engaging Active Learners
edited by Robert Lynn Canady and Michael D. Rettig

Block Scheduling
A Catalyst For Change in High Schools
by Robert Lynn Canady and Michael D. Rettig

Middle School Block Scheduling
by Robert Lynn Canady and Michael D. Rettig

The 4 X 4 Block Schedule
by J. Allen Queen and Kimberly Gaskey Isenhour

Action Research On Block Scheduling
by David Marshak

Teaching in the Block, the series
Robert Lynn Canady and Michael D. Rettig, General Editors

Supporting Students With Learning Needs in the Block
by Marcia Conti-D'Antonio, Robert Bertrando, and Joanne Eisenberger

Teaching Mathematics in the Block
by Susan Gilkey and Carla Hunt

Teaching Foreign Languages in the Block
by Deborah Blaz

Performance Assessment and Standards-Based Curricula
The Achievement Cycle
by Allan A. Glatthorn

The Performance Assessment Handbook
Volume 1: Portfolios and Socratic Seminars
Volume 2: Performances and Exhibitions
by Bil Johnson

A Collection of Performance Tasks and Rubrics

Middle School Mathematics
by Charlotte Danielson

Upper Elementary School Mathematics
by Charlotte Danielson

High School Mathematics
by Charlotte Danielson and Elizabeth Marquez

The School Portfolio
A Comprehensive Framework for School Improvement
by Victoria L. Bernhardt

Research on Educational Innovations, 2d ed.
by Arthur K. Ellis and Jeffrey T. Fouts

School-to-Work
by Arnold H. Packer and Marion W. Pines

The Reflective Supervisor
A Practical Guide for Educators
by Ray Calabrese and Sally Zepeda

Instruction and the Learning Environment
by James Keefe and John Jenkins

The Educator's Brief Guide to the Internet and the World Wide Web
by Eugene F. Provenzo

FOREWORD

Block schedules provide opportunities for teachers to change their instructional strategies so that students become more active and successful learners. There is a growing body of evidence from experiences with high school block scheduling that strongly supports the notion that with proper staff development and careful schedule design the overall school environment becomes more positive and productive. There is also evidence that many teachers increase their personal contacts with students. Furthermore, when curricular and instructional issues are addressed appropriately, achievement in many schools improves, as measured by factors such as reduced failure rates, increased number of students on honor rolls, and higher test scores.

Because we believe that instructional change is the key to successful block scheduling, we are sponsoring this series of books, written primarily by teachers who have been successful in teaching in block schedules. While we believe this series can be helpful to teachers working in any type of schedule, the ideas should be especially useful for middle and high school teachers who are "Teaching in the Block."

The idea of scheduling middle and high schools in some way other than daily, single periods is not new. We find in educational history numerous attempts to modify traditional schedules and to give the instructional school day greater flexibility. In the 1960s, for example, approximately 15% of American high schools implemented modular scheduling, which typically combined "mods" of time to create schedules with instructional periods that varied in length from between 15 minutes to classes of 100 minutes or more.

Many reasons have been given for the demise of modular scheduling as practiced during the 1960s and 1970s. However, two of the primary reasons often cited are that (1) too much independent study time was included in those schedules and school management became a problem, and (2) teachers in many schools did not receive training designed to assist them in altering instruction in the longer class period (Canady and Rettig, 1995, pp. 13–15). Current models of block scheduling do not include significant built-in independent study time and, therefore, school management problems are not exacerbated, but helped. We have found, however, that in schools where block scheduling has been implemented successfully, considerable attention has been paid to adapting instruction to maximize the potential of available time.

We repeatedly have stated that if schools merely "change their bells," block scheduling should not be implemented. We also have contended that if teachers are not provided with extensive staff development, block scheduling will be a problem. "The success or failure of the [current] block scheduling movement will be determined largely by the ability of teachers...to improve instruction. Regardless of a school's time schedule, what happens between individual teachers and students in classrooms is still most important, and simply altering the manner in which we schedule school will not ensure better instruction by teachers or increased learning by students" (Canady and Rettig, 1995, p. 240).

When block scheduling gained serious consideration in high schools, many mathematics teachers were very skeptical. They doubted that mathematics instruction could be adapted to longer periods; they expressed concern about being able to "cover" material sufficiently; and they expressed disbelief that students could take mathematics only half the year and still be ready for the next sequential course. In *Teaching Mathematics in the Block,* Susan Gilkey and Carla Hunt have addressed these and other issues related to teaching mathematics in a block schedule. They offer many ways to compact selected topics, integrate critical math concepts with other materials, and design lesson plans that maximally utilize their instructional time.

Both administrators and teachers will find their recommendations for useful resources, particularly computer software, to be very helpful. In addition, we commend their novel ideas for student assessment and for using mathematics to improve students' writing.

Robert Lynn Canady
Michael D. Rettig

About the Authors

Susan Nicodemus Gilkey, a mathematics teacher in the Albemarle County Public Schools, has taught mathematics for 22 years. She completed her masters degree at the University of Virginia, focusing on implementing the NCTM Standards, integrating reading strategies in a mathematics classroom, and using technology to teach mathematics. She has been an adjunct faculty member at the University of Virginia. With Carla Hunt, she conducts workshops and presents at conferences on implementing block scheduling in a mathematics classroom.

Carla Herndon Hunt has taught mathematics for 22 years in Texas, California, and Colorado and is currently a mathematics teacher at Albemarle High School in Virginia. She has a masters degree in mathematics education from the University of Virginia, focusing on implementing the NCTM Standards and integrating technology and graphing calculators into the curriculum. She has received numerous awards for excellence in teaching and has been an adjunct faculty member at the University of Virginia and George Mason University.

PREFACE

When we first began work on this book, we promised each other that we would write the kind of book that we ourselves would find useful. With over 40 years of teaching experience between us, we knew that this book needed to be filled with ideas, possibilities, reproducible masters, resources—anything and everything that teachers can use in their classrooms.

As our writing progressed, we realized that one book cannot do ***all*** of that. We often found ourselves exclaiming, in true Laurel and Hardy style, "Another fine mess you've gotten me into!" But we persevered. The book you hold will never be truly finished because we continue to discover more and more strategies to introduce into the classroom. But this book is a good start.

We hope that you will let us know which of the ideas and strategies in this book work for you and which ones don't. If, at first, a strategy is not completely successful, we ask that you evaluate it to determine what might make it better. Work and share ideas with your colleagues. Keep trying and growing to make your classroom a better place—you are not in this alone.

Susan Gilkey
Carla Hunt

TABLE OF CONTENTS

Acknowledgements

We want to take this time to thank the many wonderful people who aided us with this book. It is highly probable that we will miss someone. If we do, we apologize in advance. It truly will be unintentional.

We want to thank the many teachers, administrators, parents, and most of all students with whom we have worked over the years and who have knowingly or unknowingly contributed to our efforts.

Our thanks to the many members of the National Council of Teachers of Mathematics for their continuous efforts to bring about needed reform in mathematics education and their many wonderful materials. Without them, we would be greatly handicapped as teachers in our efforts to provide meaningful mathematics for our students.

At the University of Virginia, our thanks go to Dr. Joe Garofalo for thinking of us when the idea for a book on teaching mathematics in the block arose; and to Dr. Robert Lynn Canady for the many hours of discussion we had on teaching in the block, for the invaluable materials and resources he provided us during the course of writing this, and for the excellent feedback and suggestions he gave us.

Thanks to Clysta Walters, Orange County High School, for sharing her 4x4 experiences with us; Lois Williams, mathematics specialist for Albemarle County Schools, for contributing her ideas and the curriculum for middle school students; and Mary Alice Gunter, retired professor from the University of Virginia, for her inspirational teaching of instructional models and for her materials on cooperative learning.

Reviewers who offered valuable advice when this project was in the manuscript stage are Jon Arnold, St. Charles High School, MN; Patrick Cates, Lubbock High School, TX; John Dalida, Kansas State University; and Susan Socha, McLean High School, VA.

We thank Robert Sickles, our publisher, for urging us on when we got bogged down with teaching responsibilities and for appreciating the fact that we are active, classroom teachers trying to juggle an infinite number of balls at one time.

To Muffin, Lucy, and Delta for their endless patience in waiting for us to attend to their needs and to give them their treats. Good dogs! Good cat!

And last, but far from least, is our gratitude to Rollin David Larrick, a Latin, Greek, and Humanities teacher, and Sue C. Hughes, an English teacher, with whom we teach. We cannot thank them enough for the time and effort they devoted to reading our manuscript and offering invaluable suggestions to us, so that the publication draft was free of split infinitives, dangling participles, and all of those other little pesky grammar problems. They helped us in our efforts to make the book logical and readable.

DEDICATION

To Mom, who stayed with me as long as she could, encouraged me, and hopefully would be proud of me. To Marion Lewis for helping me through the troubled teen years and putting my feet on the right path. To Dorothy Overcash, my first and only algebra teacher whose position I filled when I first started teaching, for her dedication to teaching mathematics. To David, my cousin and best friend, who was there when I started teaching and who is still teaching with me now, for always encouraging and supporting me. And to Gail, my best friend, who has always been there for me.

Susan

To my Mother, a life-long educator, independent woman, and influential role model, for always making time for me and for helping me to understand the value of giving, and to my Father for telling me, "You can be whatever you want as long as you are big enough," and making me believe it. To my daughter, Jamie, and my son, Jason, for giving me so much joy and for always showing their pride in me. To Marshall, my husband Mr. Wonderful, for his encouragement, patience, and willingness to listen. And to Mrs. Bell, my fourth-grade teacher for writing on my report card, "She's a natural-born teacher."

Carla

1

NCTM Standards and Teaching Mathematics in a Block Schedule

In response to *An Agenda for Action* (National Council of Teachers of Mathematics (NCTM), 1980), *A Nation at Risk* (National Commission on Excellence in Education 1983), the National Assessment of Educational Progress (NAEP) and other reports discussing the decline in American students' ability to do mathematics, a widespread reform in mathematics education arose. The National Council of Teachers of Mathematics (NCTM) published *Curriculum and Evaluation Standards for School Mathematics* in 1989, followed by *Professional Standards for Teaching Mathematics* in 1991, and then *Assessment Standards for School Mathematics* in 1995. Their purpose was to address "the call for reform in the teaching and learning of mathematics" (NCTM, 1989, p. 1) and to put forth a vision of mathematics education for the future. All three documents underline a need for students to be active learners in constructing mathematical understanding and integrating technology when appropriate.

During this same period, educators across the country and around the world sought ways to improve educational experiences by examining ways to restructure the "traditional" middle school and secondary class period of 42–60 minutes. In *Block Scheduling, A Catalyst for Change in High Schools* (1995) and *Teaching in the Block: Strategies for Engaging Active Learners* (1996), R.L. Canady and M.D. Rettig propose alternate scheduling possibilities in which students meet in fewer classes each day for longer periods of time.

This chapter discusses the Standards from NCTM, why schools are changing to block scheduling, and the implications of teaching in a block for the NCTM Standards. The chapter provides teachers (those currently teaching in a block; those moving toward teaching in a block; those investigating teaching in a block; and preservice teachers) with a rationale for teaching mathematics in a block schedule.

NCTM Standards

We are an information society removed from the agrarian and industrialized societies of our founding fathers and ancestors. No longer are the mathematical goals of the past adequate to prepare us for the future. New societal goals must be incorporated into our educational system for all students:

- to become mathematically literate workers;
- to become lifelong learners;
- to provide mathematical opportunities for all; and
- to become members, as Thomas Jefferson said, of an informed electorate.

Education today transcends the goals of an agrarian society and must meet the demands of a technological society in which people from all parts of the world work together to find solutions for past, current, and future problems. Mathematics is valuable in this process. (NCTM, 1989, p. 3)

Societal changes necessitate changes in educational goals. Because this book deals with teaching mathematics in a block, we address the five new goals for mathematics as defined by the National Council of Teachers of Mathematics.

> Educational goals for students must reflect the importance of mathematical literacy. Toward this end, the K-12 standards articulate five general goals for all students: (1) that they learn to value mathematics, (2) that they become confident in their ability to do mathematics, (3) that they become mathematical problem solvers, (4) that they learn to communicate mathematically, and (5) that they learn to reason mathematically. These goals imply that students should be exposed to numerous and varied interrelated experiences that encourage them to value the mathematical enterprise, to develop mathematical habits of mind, and to understand and appreciate the role of mathematics in human affairs; that they should be encouraged to explore, to guess, and even to make and correct errors so that they gain confidence in their ability to solve complex problems; that they should read, write, and discuss mathematics; and that they should conjecture, test, and build arguments about a conjecture's validity....
>
> Toward this end, we see classrooms as places where interesting problems are regularly explored using important mathematical ideas.

> Our premise is that *what* a student learns depends to a great degree on *how* he or she learned it....This vision sees students studying much of the same mathematics currently taught but with quite a different emphasis; it also sees some mathematics being taught that in the past has received little emphasis in schools. (NCTM 1989, p. 5)

For many teachers, incorporating NCTM's five goals into the mathematics curriculum demands changes in mathematics classrooms. Students can no longer be viewed as vessels into which mathematical knowledge is poured; rather they must become the constructors of their own mathematical understanding. The teacher's role changes to that of a guide who assists students along their mathematical journey.

In *Professional Standards for Teaching Mathematics* , NCTM describes a "need to shift

- toward classrooms as mathematical communities—away from classrooms as simply a collection of individuals;
- toward logic and mathematical evidence as verification—away from the teacher as the sole authority for right answers;
- toward mathematical reasoning—away from merely memorizing procedures;
- toward conjecturing mathematics, inventing, and problem solving —away from an emphasis on mechanistic answer-finding;
- toward connecting mathematics, its ideas, and its applications—away from treating mathematics as a body of isolated concepts and procedures." (NCTM, 1991, p. 3)

We as teachers are instrumental in reorganizing our curriculum and classrooms to provide instructional changes that will mathematically empower our students. We must include activities that encourage students to meet the expectations of this new mathematical vision and to ensure their success as reasoning problem solvers able to communicate and to work with others. We must also construct assessment and evaluation methods matched to the learning experiences of the students.

NCTM's Evaluation Standards "proposed that

- student assessment be aligned with, and integral to, instruction;
- multiple sources of assessment information be used;
- assessment methods be appropriate for their purposes;

- all aspects of mathematical knowledge and its connections be assessed;
- instruction and curriculum be considered equally in judging the quality of a program." (NCTM, 1995, pp. 1–2)

While the traditional test still has its place in the curriculum, the test should take on a different "look." It should resemble the types of activities and thinking in which the students were engaged during the learning process and not merely be a forum for recalling facts without any application or reasoning and problem solving strategies. If students employed manipulatives or used technology in their instructional activities, the same methods should be used in evaluating student progress, understanding, and ability to do mathematics.

The *Standards* documents envision a mathematical experience that moves away from traditional instruction and evaluation toward a student's personal involvement in the learning of mathematics. Mathematics teachers across the United States and around the world are changing their mind-sets, and their teaching methods, to embrace the goals of helping all students develop confidence in becoming mathematical problem solvers who value mathematics and are able to reason and to communicate mathematically. Students who construct mathematical understanding through meaningful activities are far more likely to fulfill the NCTM goals than those who engage in rote methods with little or no opportunity to apply their learning.

WHY SCHOOLS ARE CHANGING TO BLOCK SCHEDULING

To find answers to why American students generally lag behind their foreign counterparts, educators study all aspects of the educational process, from the curriculum to the lengthening of class periods. In *Block Scheduling: A Catalyst for Change in High Schools* (1995) and *Teaching in the Block: Strategies for Engaging Active Learners* (1996), R.L. Canady and M.D. Rettig describe problems with traditional class periods and propose alternative plans in which students are in each class for longer periods of time. Their findings concerning the traditional school program in which students meet in six or seven classes everyday pointed out that:

- Instruction is fragmented for students attending schools having single period schedules.
- An impersonal, factory-like environment is created by the assembly line, single-period schedule.
- Discipline problems are exacerbated by the single-period schedule.
- Instructional possibilities are limited in short periods.

- Traditional scheduling models do not provide varying learning time for students. (Canady and Rettig, 1995, pp. 2–4)

After careful consideration of these problems, they observed and studied schools that had moved away from the traditional schedule and were using various schedules in which students met fewer classes each day and/or each semester for longer periods of time. From these studies, Canady and Rettig determined positive aspects of classes meeting in longer blocks of time and delineated that the goals of a block schedule should be to:

- Reduce the number of classes students must attend and prepare for each day and/or each term.
- Allow students variable amounts of time for learning, without lowering standards, and without punishing those who need more or less time to learn.
- Increase opportunities for some students to be accelerated.
- Reduce the number of students teachers must prepare for and interact with each day and/or term.
- Reduce the number of courses for which teachers must prepare each day and/or term.
- Reduce the fragmentation inherent in single-period schedules, a criticism that is especially pertinent to classes requiring extensive practice and/or laboratory work.
- Provide teachers with blocks of teaching time that allow and encourage the use of active teaching strategies and greater student involvement.
- Reduce the number of class changes. (Canady and Rettig, 1995, p. 6)

Most teachers agree that these are worthwhile goals, especially those that are student-oriented. If long-term assessment results demonstrate improved student learning and retention, then teaching in a block schedule will have far-reaching effects on education.

The varying schedules were examined and the benefits and concerns of the alternate-day schedule and the 4/4 semester plan were delineated. Figure 1.1 enumerates those for the alternate-day model and Figure 1.2 enumerates those for the 4/4 model.

FIGURE 1.1. ALTERNATE-DAY SCHEDULE

Benefits of an Alternate-Day Schedule

- Teachers benefit from increased instructional time.
- Teachers are able to plan lessons for extended periods of time.
- The number of class changes is reduced.
- Because one or more days lapse between classes, when discipline problems occur both teacher and students have time to "cool down" before facing each other again.
- Compared to single-period, daily schedule models, students in an alternate-day schedule have fewer classes, quizzes, tests, and homework assignments on any one day.

Alternate-Day Concerns

- Maintaining students' attention
- Providing balanced teacher planning time
- Providing balanced daily schedules for students
- Providing a predictable calendar for planning school activities
- Requiring additional review time

(Canady & Rettig, 1995, pp. 8, 9–10)

FIGURE 1.2. 4/4 SEMESTER PLAN

4/4 Semester Plan Advantages

- Teachers work with fewer students during any one semester.
- Teachers prepare for fewer courses each semester.
- Teachers generally have longer and more useful planning.
- Teachers and students have two "fresh starts" each year.
- Teachers must keep records and grades for only 60–90 students per semester.
- Students concentrate on only four courses per semester.
- Students may retake failed courses in the second semester.
- Eight credits can be earned without the stress of taking eight courses at the same time.
- Fewer textbooks are required.

4/4 Semester Plan Concerns

- Will retention of learning decrease?
- Will we be able to cover the same curriculum?
- How will we handle AP courses such as U.S. history or calculus?

- How will students participate in programs such as band, choir, and orchestra which must meet both semesters?

(Canady & Rettig, 1995, pp. 13–14)

In both of these block schedules, the benefits have more implications for teachers than students, while concerns center around the students. Our purpose is not to argue for or against any schedule, but to point out that the advantages and concerns of one plan can be found to a certain degree within the other plan. For example, keeping a student's attention applies to any schedule, especially those which increase the length of the class. Also, retention problems, even in classes which meet daily for 50 minutes, plagues teachers.

In addition, each school must be able to delineate its specific reasons for changing to a block schedule, keeping students and curriculum at the top of the list. While the ideal schedule provides a planning period every day for teachers, it is not always possible as it is in the alternate six-period day schedule. However, teachers accomplish more in one long planning period than in two short planning periods: administrative responsibilities can be addressed; telephone calls dealing with school business can be returned; and there is still time to design lessons, activities, and assessments.

Even though block scheduling presents teachers with curricular and instructional challenges, especially in the initial stages, the majority of teachers are favorable toward block scheduling after experiencing it for two or more years. Students also are reported to have positive feelings toward block scheduling (Angola High School, 1996; Einder, 1996; Guskey & Kifer, 1994; Hundley, 1996; Jones, 1997; King, Clements, Enns, Lockerbie, & Warren, 1975; Pulaski County High School, 1994; Sessoms, 1995).

The next section addresses how block scheduling aided our implementation of the *Standards* in our mathematics classrooms. Subsequent chapters provide techniques for teaching in a block; planning; instructional strategies; using technology; assessment; and activities and lesson plans that have been used in longer class periods.

BLOCK SCHEDULING AND THE NCTM STANDARDS

The *Standards* advocates an active classroom in which students are "busy" doing mathematics. Teachers become facilitators for the students, increasing emphasis on mathematical connections and on mathematics as problem solving, as communication, and as reasoning. Instructional practices must change, employing those strategies that actively involve students in "a variety of instructional formats (small groups, individual explorations, peer instruction, whole-class discussions, project work" (NCTM 1989, p. 128), "exploring, conjecturing, analyzing, and applying mathematics in both a mathematical and a real-world context." (NCTM, 1989, p. 72) Students become the architects of their mathematical knowledge. Also, schools must make appropriate technology available and provide training opportunities so that teachers and students may become proficient in using the technology for exploring and doing mathematics. Teachers must find ways for "the systematic maintenance of student learning and embedding review in the context of new mathematics and problem situations." (NCTM, 1989, p. 72)

In a traditional class period, time must be allocated in such a way that all of these things can happen. The difficulty we found with the traditional, shorter class period is that there is not enough time to provide as many quality mathematical experiences for students as in longer classes. A typical scenario rushes the class: going over homework, introducing new material and having time to practice new concepts and skills before the next homework assignment is given. This becomes a repetitive cycle in which students have more and more questions about homework problems because there was not enough time in class to develop a full understanding of the new material, preparing them for the homework. Valuable instructional time is consequently lost and never recovered, so parts of the curriculum are either eliminated or slighted. It becomes nearly impossible for students to engage in real-world problem situations let alone to participate in laboratory experiments.

In a block schedule, longer periods of time afford teachers and students opportunities to engage in meaningful mathematics. Teachers may design an increased number of "worthwhile mathematical tasks" that are based on "sound and significant mathematics...that engage students' intellect; develop students' mathematical understanding and skills;...represent mathematics as an ongoing human activity;...promote the development of all students' dispositions to do mathematics." (NCTM, 1991, p. 25)

Block scheduling provides mathematics teachers more freedom "to create a learning environment that fosters the development of each student's mathematical power by providing and structuring the time necessary to explore sound mathematics and grapple with significant ideas and problems; using the physical space and materials in ways that facilitate students' learning of mathematics;...and by consistently expecting and encouraging students to work independently and collaboratively to make sense of mathematics." (NCTM, 1991, p. 57) Field trips and outdoor activities may become a reality instead of just a desire; students break away from the chalkboard to go out to the field to do mathematics.

DEALING WITH THE CHALLENGES

Teachers ask questions about how to deal with various challenges in block scheduling: how to deal with retention issues; how to ensure success for "slow learners"; how to teach certain mathematical subjects, especially middle school courses and Algebra I; how to cover the appropriate material in AP Calculus and AP Statistics; and what to do about students who need to repeat a course. While most of the answers are similar to those for teaching the mathematics curriculum during a traditional class period, modifications must be made for some of the block schedules.

RETENTION

Retention, "slow learners," and repeaters are issues with which all teachers must deal. If there were "tried and true" answers, all teachers would have success with these students and the questions pertaining to them would be moot. In the block, review of concepts and skills must be built into the curriculum to accommodate those students who need more experiences to construct their mathematical framework and to find ways to bring back prior learning so that it may be applied to problems both old and new. Not only will those students who need more time and more practice benefit but also those students who already understand will find another opportunity to apply their mathematical learning.

MIDDLE SCHOOL AND ALGEBRA I

Middle school classes and Algebra I classes should be taught with a mind-set that the more opportunities students have to actively construct their own learning of mathematics, the better the chances are that they will take ownership of that learning. However, this mind-set is true of all mathematical experiences and, hence, the process of embedding review becomes important. Specific ideas for Algebra I are outlined in the two chapters on planning; instructional strategies are discussed in Chapter 4. By using strategies from your own repertoire, and ideas within these specific chapters and other parts of the book, you will be able to find ways to ensure retention and success for both groups of students.

AP COURSES IN A 4/4 PLAN

The 4/4 semester plan poses unique problems for Advanced Placement (AP) courses. If the course is taught in the first semester, retention for the test in May for a course that concluded in January may present difficulties. For the course taught during the second semester, problems arise because in some schools, by the first part of May, up to one-fourth of the curriculum is yet to be covered, so students are not prepared for all of the questions they will be asked. The answer is either to teach the course in a year-long format and offer additional credit or to tailor it to 27 weeks and plan an elective course for the remainder of the year. Traditional AP courses cannot successfully be taught, in preparation for an AP Examination, in the 4/4 semester plan without modification. Also, the Educational Testing Service (ETS), which publishes the AP Examinations, has no plan to administer the test other than once a year in May.

COURSE REPETITION

The problem of students needing to repeat a course is more easily handled in the 4/4 semester plan than in any of the other blocks, in that students may immediately repeat the course the following semester. Immediate repetition of courses eliminates electives, but seems to ensure a higher rate of student success in subsequent courses. Repetition of a course in an alternate-day plan provides nearly the same outcome as in a traditional schedule.

CONCLUSION

Block schedules facilitate the implementation of the *Standards* in their fullest form. Both advocate the cultivation of mathematical learning in an active environment that encourages the use of varied instructional strategies and integrates appropriate technology and tools. Excellent opportunities exist in a block schedule for embedding review within new material so that retention occurs and learning is placed in long-term memory for future retrieval and application. Teaching in longer blocks of time optimizes the possibilities for the realization of the goals and "standards" of the National Council of Teachers of Mathematics.

Changes need to occur in curriculum, instruction, and assessment to be successful in meeting the *Standards*.

These same changes need to occur for teaching in a block schedule. The shifts below, identified by NCTM to ensure mathematical learning, parallel changes which need to occur in a block schedule if students are to meet with success.

- Shift in content:
  Toward a rich variety of mathematical topics and problem situations
  Away from just arithmetic
- Shift in learning:
  Toward investigating problems
  Away from memorizing and repeating

- Shift in teaching:
  Toward questioning and listening
  Away from telling
- Shift in evaluation:
  Toward evidence from several sources judged by teachers
  Away from a single test judged externally
- Shift in expectations:
  Toward using concepts and procedures to solve problems
  Away from just mastering isolated concepts and procedures

(NCTM, 1995, pp. 2–3)

Teaching in a block schedule provides an ideal time for planning and teaching by weaving a tapestry of concepts together. A conceptual framework provides opportunities for investigations, explorations and discovery in a context of meaningful problems employing current technology in a cooperative learning environment. Students learn to value mathematics which is not completed in isolation.

Students' assessments should incorporate the strategies teachers used while teaching the concepts, including both group and individual evaluations. Developing multiple assessment methods to gather information about student learning furnishes students, parents, and teachers with a better picture of student progress. Block scheduling provides the time and opportunity to accomplish this.

2

Curriculum Planning and Instructional Design for the Block Schedule

"How do teachers decide what to teach? Once they make a decision, how do they turn their topic into a lesson that will work for school kids? Watching teachers in movies or on TV or even in school classrooms provides few clues about prior planning. Teachers just seem to do it. Yet in truth you can't walk into a room full of students and just begin, even if you're loaded with knowledge, ideas, energy, and good will. Classroom teaching doesn't work like that. Although it hardly ever seems so to those who haven't done it, it is one of the more complicated tasks on earth. You must plan.

Teachers are inundated with guidelines directing their curricular decisions. These range from state-mandated course curricula to textbooks so teacher-proof that they require minimum thought and creativity. I believe that teachers have more control over what they teach than they realize. The most important of the many frameworks guiding teachers in the classroom should be self-determined and self-imposed. Otherwise, what's the point of being there with the kids? Making responsible decisions about what to teach, understanding why you are teaching it, and determining the best way for the students to learn it are part and parcel of the work of the professional teacher."

David Kobrin, *In There with the Kids* (1992, p. 47)

Planning What to Teach

Teachers know that mathematical success in the classroom depends upon the many hours of work and thought that take place before the bell rings. Not only do they need to plan what should be taught but they also need to determine how it should be taught. Choosing appropriate activities enables students to become active learners. Finding meaningful ways to ensure that learning actually occurs, and then ensuring that the learning is not only remembered but retained, is crucial.

In a block schedule, this is even more critical. Not only are teachers faced with keeping students' attention focused for longer periods of time, but they must also plan with the knowledge that they will not see these students as many times as they do during a traditional schedule in which the class meets daily. While this may seem like an insurmountable obstacle at times, planning with it in mind makes the challenge less overwhelming.

This chapter addresses curriculum compacting and offers suggestions for managing problems that may be encountered with the challenge. It also offers ideas for regarding teaching in a block as a rewarding experience for both teachers and students, allowing for learning experiences that were previously not possible in shorter periods of time. Finally, there are ideas that should make the changes a teacher needs to make for teaching in a block more manageable and less stressful.

CURRICULUM REORGANIZATION

Whether your school has adopted an alternate-day plan, a 4/4 semester plan, a 75-75-30 plan, or some variation of one of these schedules, restructuring what you are going to teach is essential. You cannot meet with success by stacking two or three lessons, one on top of the other. Lessons will not build smoothly one on the other and learning will become fragmented. While stacking may seem like an adequate plan for compaction, students will not benefit from it. Goals and objectives will become indiscernible to teacher and learner alike.

Neither can you stretch one lesson over an entire block, one per day, and ever hope to complete your curriculum. Even though spending so much extra time on one lesson from your traditional plans may improve retention of the specific objective, you will lose valuable and necessary parts of your curriculum and students will suffer over time.

What is the solution? You must build a framework for your curriculum that allows for completing more of the course in one block while ensuring learning and retention of the material. Integrating concepts, skills, and processes together creates a tapestry to hold your course framework together. Changing the structure of the curriculum also affords an opportunity for incorporating problem solving, reasoning, communication, and mathematical connections in a meaningful way. Making provisions for these to occur in the traditional shorter class period is nearly impossible because the available time must be spent "delivering" new information.

With continued positive emphasis by the NCTM on making mathematics concept-driven and student-centered, the block can indeed become your friend and ally. It can afford you the opportunity to become "the guide on the side," a role Michael Serra advocates, as opposed to "the sage on the stage." With careful planning, you can stand back and really "see" and "hear" learning occur as students engage in mathematical activities. Waiting until you give a quiz or a test to see if what you were teaching really was what the students were learning is too long, especially in a block.

In a 4/4 semester model or a 75-75-30 model, you need to arrange units so students will have the best opportunity for retaining the concepts, processes, and skills from course to course, especially because many students may not take another mathematics course for up to a year. In the alternate-day model, building retention from class to class becomes an issue, but is not as difficult as ensuring remembrance over longer periods of time. Teachers have always been faced with retention problems over the passing of a summer and somehow have found ways to adjust and bring students' prior learning back. Even memory research shows that after an initial amount of "forgetting" occurs very little information is subsequently lost.

The core of any curriculum should **not** change, no matter which block schedule you use. The curriculum is all important and cutting out crucial elements will not create students with better mathematical understanding and more mathematical power. On the contrary, just the opposite will happen. They will be left with incomplete frameworks of understanding, and the missing components will leave them at a disadvantage for future learning.

TAKING THE FIRST STEP

To begin the process of restructuring your curriculum, find a colleague who is teaching the same course as you and work together. Having another person to use as a sounding board and to share in the planning gives you the support necessary to tackle the task. Most times "two heads are better than one," and working with someone else makes the process easier and doubles the information and resources available for building this "new" course.

If your mathematics department is large enough, form teams of teachers by course to plan for teaching the curriculum, to find ways for incorporating activities into each lesson, and to design meaningful assessments. Each member of the team will bring a variety of ideas as well as activities and teaching strategies to the group. Finding the lessons, activities, and strategies that worked in the classroom previously alleviates the Herculean task of curriculum restructuring. Restructuring does not mean throwing out everything from "before block," but rather finding the best of the "tried and true" methods and building upon, adding to, and reincorporating them into the new time frame.

Examine the textbook and be willing to break away from the tendency to follow its order unit by unit and lesson by lesson. The textbook was written by people who determined that its order might work best for them or because many of the textbooks were following a similar format. Even though most texts will probably closely align to your curriculum, the one you are using might not be the best for you and your schedule. More importantly, its order might not be best for your students. You need to be flexible and be willing to change and to look for ways to ensure that mathematical learning occurs and is retained.

Change needs to occur in the way you arrange the teaching of units on the macro level and the ways you teach individual lessons on the micro level. The goals of teaching in a block should not include a plan for eliminating curriculum but rather for finding methods that will incorporate all of the concepts in a way for the course to become stronger.

In any schedule, you should plan to include at least half of the units during the first half of the teaching days: in an alternate-day plan, during the first semester; on a 4/4 plan, during the first 45 days; and on a 75-75-30 plan, during the first 35 days. Adjustments can be made for those schedules that are variations of these. This allows you time to monitor and adjust as you move from concept to concept and through the units. The willingness and ability to adjust as you go is critical for the success of your students.

Also, look for ways to "circle back" and revisit topics. This can be accomplished by beginning the class with "warm-ups" in which the students are given three to five problems covering previous concepts, by including a few problems from previous concepts in homework assignments, and through classroom activities that incorporate previous concepts with new ones, as well as by making quizzes and tests comprehensive. The more often students are exposed to concepts, the better they become at retaining them, and, more importantly, at understanding and at applying them at a later time in a new context. Building in specific activities for this becomes easier with practice and offers students increased opportunities for problem solving, reasoning, communication, and connections.

WHAT'S ALL THE FUSS? ALGEBRA I AND BLOCK SCHEDULING

Many teachers express concern about teaching Algebra I in the longer class periods which meet fewer times. Those teachers who are on a schedule that is not year-long and who teach the so-called "lower level students," that is, those students who may not particularly like mathematics and/or who have not always experienced success in it, find the block schedule an overwhelming challenge at times. They say that they are unable to complete the curriculum and/or that students do not meet with success.

These concerns cannot be ignored; they must be addressed and ideas generated so both teachers and students enjoy their mathematics classes. Also, and more importantly, methods should be identified by which the students are able to meet with success in Algebra I and future courses while teachers complete the curriculum. Figure 2.1 is the table of contents from an Algebra I textbook. Figure 2.2 is a rearrangement of the chapters into teaching units.

FIGURE 2.1. *ALGEBRA I* TABLE OF CONTENTS

1 Connections to Algebra
2 Rules of Algebra
3 Solving Linear Equations
4 Graphing Linear Equations
5 Writing Linear Equations
6 Solving and Graphing Linear Equations
7 Solving Systems of Linear Equations
8 Powers and Exponents
9 Quadratic Equations
10 Polynomials and Factoring
11 Using Proportions and Rational Equations
12 Functions
13 Radicals and More Connections to Geometry

Larson, Kanold, & Stiff, *Algebra I* (D.C. Heath and Company, 1993)

FIGURE 2.2. REARRANGEMENT OF *ALGEBRA I* TABLE OF CONTENTS INTO TEACHING UNITS

First Semester

I. Linking Arithmetic to Algebra
II. Solving Linear Equations
III. Matrices
IV. Graphing Linear Functions and Relations
V. Graphing Absolute Value
VI. Writing Equations of Linear Functions
VII. Solving and Graphing Linear Equations and Inequalities
VIII. Solving Systems of Linear Equations and Inequalities

Second Semester

I. Powers and Exponents
II. Polynomials and Factoring
III. Quadratic Functions
IV. Proportions and Variation
V. Rational Expression and Equations
VI. Data Representation and Analysis
VII. Radicals

Careful planning for curriculum realignment points toward a solution for both completion of the curriculum and for student success and retention. Redesigning the algebra objectives in a way that students frequently return to them and apply them in new contexts ensures building the bridges and connections necessary to construct a strong algebraic foundation. Incorporating several concepts that are closely associated eliminates teaching individual objectives in isolation from each other and helps students make necessary connections.

For example, all of the concepts in the first three chapters of the text are concepts from prealgebra which should have been completed prior to Algebra I. Rearranging them into two units, in which students review and master their concepts through hands-on activities while working cooperatively, not only decreases the time a teacher must devote to the prealgebra skills and concepts but also affords an opportunity for the teacher to determine which students need more reinforcement and which students are algebra-ready. A further example is given in the next section.

Teachers recognize that a strong foundation in the skills and concepts found in Algebra I is essential for undertaking future mathematics courses. Working together and seeking, finding, and implementing solutions to problems may alleviate not only difficulties encountered in the Algebra I curriculum but also in other mathematics courses.

ONE SCHOOL'S SOLUTION

Orange County High School, in Orange, Virginia, has built special accommodations into its 4/4 semester plan. All eighth graders not taking algebra are tested for algebra readiness. The students whose scores show readiness register for Algebra I in the fall of their ninth grade year. Those students who are not prepared enroll in an elective course for nonmathematics credit to prepare them for Algebra I. At its conclusion, these students immediately take Algebra I. Students taking Algebra I must complete it with a grade of C or better to continue to Geometry and Algebra II. If they do not, they take a course in elementary functions, which reteaches and reinforces the skills and concepts from the Algebra I curriculum. Successful completion of this course allows students to enroll for their next mathematics class. Many students at Orange High School successfully complete all of the course offerings and enroll for additional mathematics courses at the local community college. Most students graduate with more than the minimum mathematics requirements.

Even though the Algebra I curriculum proves to be a great challenge for some teachers and students, there are many ways to incorporate strategies for success and retention within the course framework. Remaining flexible and working with colleagues to develop appropriate methods should ease any frustrations.

ONE COURSE AND ONE POSSIBLE WAY: A MACRO VIEW

Let's look at a typical Algebra II course. In a year, each of us probably includes a brief review of Algebra I material followed by units on linear and quadratic functions; matrix operations; functions in general, and polynomial, exponential, logarithmic, and rational functions; the conic sections; and sequences and series. This is the core curriculum; it is not optional nor should pieces of its body be truncated in order to include all of its parts. What can we do with this imposing set of concepts, processes, and skills so that the Algebra II curriculum is complete as well as rich, rewarding, and retained?

For an Algebra II course, we take the textbook, rearrange its units, and teach the first unit, a review, with the inclusion of new concepts. Among the additions are activities to incorporate unit analysis, which reinforces concepts from science courses. It provides for a better beginning and reinforces previously learned concepts in a novel way. Students stop viewing the beginning of the new course as, "Been there. Done this. Ho hum. I'm bored!" It forces them to rethink their previous paradigm that all mathematics courses spend much of the first semester rehashing topics from the previous course. Immediately, they must focus on the business of mathematics.

Next, we approach linear functions using materials that put the students at the center of learning. Incorporating laboratory activities that involve them in small groups introduces cooperative learning, which raises an awareness that mathematics need not be an endeavor by individuals in isolation. Some of them may remember concepts about lines from Algebra I and Geometry, and those who do not can work with their peers to refresh their memories in a short time and work toward mastery of the concepts.

In the linear function unit, introducing the concept of determining the equation for a linear function by using finite differences and difference tables provides a change of pace. Students appreciate the abstraction that can be developed from concrete tables and the concept can be reintroduced to extend the process and concept to quadratic and higher degree functions in later units.

From here a natural flow takes you to linear systems and into matrices for solving more complicated systems of linear equations and applications. These units can be enriched by including many real-world applications and examples, especially problems from linear programming that minimize cost or maximize profit.

Moving from linear experiences to quadratic functions follows logically. In this unit, factoring is reintroduced to solve certain types of quadratic equations. It is not necessary to wait for it to pop-up in the unit on polynomials, especially because the students usually need to use it long before it appears in the textbook. You can revisit it in the polynomial unit for factoring other types of equations, especially those of higher degree or those that pattern themselves after the quadratic types.

Conic sections are fundamental to many later mathematics courses. Since they normally fall near the end of the course, and many teachers are rushed just to get this unit in, why not move the unit with these quadratic relations to immediately follow the unit on quadratic functions? Factoring and completing the square concepts are in place. Students make the link between parabolas and their other quadratic cousins more quickly than waiting until near the course end. The conic applications are rich and varied. Introducing analytical thinking sooner in the course benefits students. Also, it reinforces that Algebra II is not a rehash of Algebra I topics with a few new ideas thrown in toward the end of the course.

This should bring you to "mid-year" and the first semester exam. It affords an opportunity to conduct an inventory on how things are proceeding and what needs to be adjusted for the new semester. Ask yourself these questions: Has half of the curriculum been completed? Do the students understand the skills and concepts, and can they apply them? Have enough activities been incorporated into the lessons? Has the pace been correct?

Beginning the "second semester" with a more in-depth study of functions and their operations allows students the opportunity to "revisit" familiar concepts and to build on them through composition and recursion. Next a unit on exponential functions succeeded by their inverses, the logarithmic functions, provides a way to weave more concepts together. Polynomial and rational functions follow and the "year" concludes with a unit on sequences and series.

If time remains at the end of the course, topics can be revisited, extended, or enriched by integrating more applications from the real world, problem-solving activities, or projects to foster communication skills and make connections to other disciplines or branches of mathematics.

As is evident from this plan, you should not wait until the middle of the course to reintroduce the notion of a function from Algebra I. There is no reason to separate the concept from the linear experience, especially because functions are not new to students. Middle school classes, and even elementary school classes, introduce and deal with an elementary concept of functions. The science curricula are function-driven. Teaching skills and incorporating concepts when and where they are appropriate and applicable, not just where an author has placed them in the text, builds stronger and more meaningful understanding.

The textbook should serve as a resource, not as the curriculum. As long as the bridges from concept to concept are in place and the students are able to make connections between concepts and previous mathematical experiences, it's best to teach the curriculum in the way it makes sense to you. If the links are not readily apparent to the students, help them find the relationships. If they are unable to see the connections, let alone make them themselves, retention is minimal at best.

A UNIT: ONE POSSIBLE WAY—MOVING AWAY FROM MACRO

Let's take an Algebra I unit and give it some definition in terms of objectives and activities. Again, look at the order a typical textbook uses and then investigate ideas for integrating the objectives into a unit framework which will encourage meaningful student learning and retention.

In Figure 2.3 are the "What you should learn" goals from Chapter 4, "Graphing Linear Equations," in *Algebra I* by Larson, Kanold, and Stiff (D.C. Heath and Company, 1993). Figure 2.4 rearranges and enriches the same goals.

FIGURE 2.3. *ALGEBRA I* "WHAT YOU SHOULD LEARN" GOALS

1. graphing horizontal and vertical lines in the coordinate plane
2. using equations of horizontal and vertical lines in real-life settings
3. graphing a linear equation from a table of values
4. interpreting graphs of linear equations
5. finding the intercepts of the graph of a linear equation
6. using intercepts to sketch a quick graph of a line
7. finding the slope of a line using two of its points
8. interpreting slope as a constate rate of change or as an average rate of change
9. finding the slope and y-intercept from an equation
10. using slope-intercept form to sketch a line and to solve real-life problems
11. using a graph as a quick check of a solution found algebraically
12. approximating solutions to real-life problems by using a graph

FIGURE 2.4. "WHAT YOU SHOULD LEARN" GOALS REARRANGED AND ENHANCED

1. graphing lines in one variable (horizontal and vertical lines)
2. graphing lines in two variables
3. graphing lines using x- and y- intercepts (quick graphing)
4. finding the slope of a line
5. graphing using slope-intercept form of a line
6. solving linear equations in one variable and x-intercepts

7. applying and connecting to the real world
8. determining the difference between a relation and a function
9. using laboratory activities to make connections among algebraic concepts, to other disciplines, and to technology

The textbook goals incorporate the necessary skills and concepts that should be included in an Algebra I unit on graphing linear equations. In the block, a rearrangement and enrichment of these goals lets the teacher use the text as a resource for reading and for assigning exercises and problems. It also allows flexibility in ordering the text topics and provides opportunities for incorporating additional topics appropriate for this unit.

Designing this unit to incorporate algebraic skills and concepts entails determining which concepts should be taught together, the strategies and activities that should be used to teach them, the time that must be spent on the unit, and the number of lessons and amount of time that must be spent on each lesson. The topics can be arranged for the enriched course into five lessons, with a sixth class for review and assessment. Figure 2.5 is the rough outline of the topics for the first lesson for this unit with a few ideas for teaching the lesson.

FIGURE 2.5. TOPIC OUTLINE FOR LESSON ONE

Lesson One

Use direct instruction and questioning to review graphing in the coordinate plane.

Have students graph a simple picture on a coordinate plane.

Develop the concept of function using concept attainment.

Use direct instruction function notation, domain and range.

Have students graph coordinate pairs from tables.

This outline is in skeletal form and can be used to develop a complete lesson including objectives, materials, instructional strategies, activities, guided practice, independent practice/homework, extensions, and assessment. This lesson incorporates a review of prerequisite knowledge about the coordinate plane, reintroduces the concept of function with its appropriate vocabulary, and includes graphing coordinate pairs from a student-generated table of values which reviews the process of evaluating expressions.

CONCLUSIONS

Planning is key to the success of teaching in a block. In the beginning, sizable quantities of time must be devoted to realigning the curriculum as well as to monitoring and adjusting to ensure clarity, continuity, and coverage.

Teachers should not eliminate critical parts of the curriculum just to "cover it all." Rather they should reorganize the curriculum into units that may be enhanced as a result of teaching in a block.

Working together with colleagues minimizes "burning the candle at both ends." Plans and ideas are doubled with half of the time committed. Brainstorming and sharing helps troubleshoot difficult areas and makes both of you feel like your are "not in it all alone."

3

Lesson Planning and Instructional Design for the Block Schedule

Deciding How to Use the Block

Once restructuring the curriculum and designing the units to fit into a given block model are completed, decisions must be made concerning how to allot the time during each class. Mathematics learning should not be static; it should be alive and energetic with students actively participating in the steps toward mastery of its many concepts, skills, and processes while constructing an understanding of the discipline itself.

Planning the lessons that make up a unit cannot be separated from planning the unit. The lessons are the threads of the tapestry of mathematical understanding and must be woven together to ensure both strength and clarity. While some direct instruction is necessary, using it for an entire 85- to 100-minute class portends disaster. Activities must be rich and varied and must employ strategies that enable the student to focus.

This chapter addresses several plans for allocating time in a block schedule. The chapter on instructional strategies discusses individual strategies for teaching in a block.

Allotting the Time and Sample Designs

Allotting time during a block varies from block to block, course to course, and teacher to teacher. How you choose to use the time for an individual lesson depends upon both the objectives of the lesson and the activities you have chosen for teaching those objectives, as well as on the type of learning environment you want for your classroom.

The teacher who favors a less noisy and more orderly environment will prefer a plan that advocates a more structured lesson design centered around individual activity and with students in their seats much of the time. On the other hand, the teacher whose style thrives in a group-dynamic, hands-on, activity-oriented class will lean toward a lesson design that encourages cooperative learning with students up and around the room engaged in a laboratory-type setting. Both of these mathematics teachers can meet success in a block setting as can those teachers whose style falls somewhere between the two.

One design will not satisfy all lessons. You must be flexible and willing to adjust the learning environment based on the curriculum being taught. Experiment to find which design works best in a given situation. Be flexible and willing to monitor and adjust as you would in a traditional 50-minute period. Although some teachers think it ideal to have a single lesson template that works all of the time, for all classes and for all students, such a plan is impossible to construct and would likely ensure monotony and failure both for teacher and students alike.

Critical to the success of block teaching is varying teaching strategies within the period. Changing activities several times during a block prevents student and teacher boredom. A rule of thumb is to strive for three changes to keep the class alive. Changing activities promotes greater mathematical understanding, encourages incorporating real-world applications and mathematical connections, and gives the teacher an opportunity to build-in reinforcement activities.

The change can be from something as simple as moving from lecturing to having students do some exercises for reinforcement of a new skill or concept. The chapter on instructional strategies provides many ideas for varying instruction; Appendix B, on resources, provides the names and locations of materials to use in your classroom.

A More Traditional Class Schedule

The first lesson design (Figure 3.1) is the type usually found in a more traditional 50-minute class; however, it has been adjusted for the longer class time.

Beginning each class with some type of warm-up activity sets the tone for the day. Students immediately focus on and engage in the business of mathematics. It gives teachers time to take roll or to take care of administrative chores, to circle around the room to speak with students one-on-one, to discuss make-up work, or to check who has and who has not completed the homework assignment, without wasting class time. Getting students started as soon as they walk in the door helps them see that you view the time you have with them as important and not for wasting.

FIGURE 3.1. TRADITIONAL LESSON DESIGN

Type of Class Activity	Minutes
Warm-up/Problem-solving	10–15
Writing	5–10
Questions on Homework	10
New Material	25–30
Practice Activity	15
Closure	10
Indpendent Practice	10
Total Time	**85–100**

There are many different types of warm-ups:

- Use four 4s and any mathematical operations, including the square root function, or use a number as an exponent to write the numbers from 1 to 10.
- How many pieces of pizza can you get if you cut the pizza six times? Cuts do not have to go through the center of the pizza.
- How many people do you have to have in a room to guarantee that at least four of them share the same month of birth?
- Use one or more problems to review the previous assignment or to reinforce previous skills or concepts.

The chapter on instructional strategies gives many examples. The list of possibilities is nearly endless. The monthly calendar in NCTM's *Mathematics Teacher* provides excellent problems. The ways to begin class are many and varied; the importance of any selection is to focus students on mathematics.

WARM-UPS

As in lesson design, warm-up activities may vary from day to day and lesson to lesson; however, establishing a routine of warm-up activities is beneficial. The activity should be carefully chosen to enhance the mathematical understanding of the students.

In the case of a novel problem, such as how many squares are there on a chess board, students may not arrive at an answer in the time you have allotted to the task. Also, they may not want the answer immediately. Wait and provide it to them individually when asked, or go back to it at the end of the class or at the beginning of the next class.

Invite students to bring in ideas or problems to share with the class. The students can write them on an overhead transparency or, if any are too long or complex, have them prepared ahead of time on an overhead transparency. Have the student who brought the problem in present it to the class as you would present it. If you want to build a bank of problems and at the same time encourage student participation, offer points for student-contributed problems in a way that would be consistent with your grading system. Contributions could be for extra credit or incorporated into an assignment.

WRITING

Writing can be incorporated into the warm-up activity or included as a separate part. It could be used for expository writing to explain a concept or algorithm. Detailed examples may be found in the chapter on instructional strategies. Writing to summarize the main concepts of a lecture or activity is an effective way to end a class. Not only does it serve to leave the student with a completed idea of mathematical content, but it also affords the teacher an informal opportunity to gauge student understanding.

CHECKING HOMEWORK

"Checking" homework presents difficulties in the 50-minute class, and can "eat up" time during a block as well. You first must determine your reason for checking student homework. Do you want to know that the student tried the assignment and gained practice and mastery of a concept, skill, or process? Do you want to know if the answers are correct? Once you decide your purpose, choosing a method makes sense.

Accepting homework that only has the answers and no accompanying work may send a confusing message to students. It may tell them that your emphasis is on correct answers rather than on the process or on the thinking that went into arriving at an answer. Also, it may contribute to the mind-set that there is one, and only one, right answer in mathematics.

The best way to keep the whole class involved is to conduct whole-class discussions about the problems that many of the students ask about. Questions about a particular exercise posed by only a few students can be answered on an individual basis sometime during the block. If there are no questions, creating another question on the concept or discussing a particular problem from the homework may provide students with an understanding they had not grasped when they originally attacked the problem.

NEW MATERIAL

Presenting new material can be done in multiple ways. Sometimes direct instruction is not only the most efficient but the least likely to create misconceptions. However, a lecture on new concepts and ideas every day brings monotony to the class for you and your students. Varying presentation strategies energizes the learning process.

Students should practice the new concept, skill, or process to ensure their immediate understanding. They should practice in the classroom with you and their peers, either individually or in small groups, to gain facility in meshing the new material with the old. No great writer ever received acclaim for just reading the words of another writer. Worthwhile writing takes first drafts, revisions, second drafts, and more revisions, until the writer is satisfied that the work is the best possible. Just as writing practice makes writers better, mathematics is also a discipline that requires practice. Greater student understanding can be achieved by providing many opportunities for students to work problems that reinforce concepts during class. Then homework can incorporate applications and connections.

Closure should be an automatic part of every class. Not only will it bring a smooth ending to the class, but it will also leave the students with another perspective on the new material and provide the teacher an opportunity for informal assessment of student understanding.

Independent practice (homework) can be assigned at this point if students need a detailed and careful explanation or clarification of its directions. It is best not to automatically build-in class time to begin homework every day because students will waste the time. An extra problem could be posed to engage the students, or a question could be posed for a short writing assignment.

A SCHEDULE FOR GROUP WORK

Figure 3.2 presents another time management plan, similar in many ways to our first model. It differs in that it builds in specific times for students to work together. Group work lends itself to discovery learning. Students first receive some information and then work in small groups to master material. Then they come back together and present the findings from the individual groups to the entire class. Finally, there is a large group interaction followed by some type of closure activity.

This plan would support many of the different instructional models, including concept development, concept attainment, inquiry, jigsaw, and TGT (teams, games, tournaments), which are discussed at length in the chapter on instructional strategies.

FIGURE 3.2. LESSON PLAN WITH GROUP WORK

Type of Class Activity	Minutes
Review/Focus/Warm-up	10
Lecture/Facts	10–15
Small Group Work	20–25
Small Group Presentations	20–25
Large Group Interaction	15
Closure/Assignment/Homework	10
Total Time	**85–100**

A LABORATORY CLASS SCHEDULE

We designed the next time management plan (Figure 3.3) to accommodate the integration of laboratory activities into the mathematics curriculum. Laboratory lessons can be as simple or as complex as you wish. They do not have to involve a great deal of equipment, and are ideal if used to collect and analyze data. Several lab activities are included in Appendix A. There are many laboratory books available commercially that include objectives, complete descriptions for conducting the labs, and lab worksheets. Some include the use of technology such as the graphing calculator and/or CBL (calculator-based laboratory). Several different labs can be designed around a bag of M&Ms (from middle school measures of central tendency through secondary investigation of exponential decay functions). Also, working with a science teacher to find appropriate activities adds an interdisciplinary connection to the lesson design, and reinforces the importance of mathematics to other disciplines.

Choosing which labs to use is important. Each should be incorporated within the curriculum objectives and have definite purpose and meaning. Conducting a lab just to vary class structure is not sound. Prelab activities explaining the purpose of the lab and its relationship to the particular mathematics concept are essential for providing students with an appropriate framework for the mathematics you want them to learn. After the data collection and analysis, it is equally important to wrap up the lab with a whole-class discussion that relates mathematical concepts to the lab activity and makes certain that students recognize and understand the connections between mathematics and science.

Included in Appendix A on lesson plans are several complete labs ready for classroom use, with introductory and closing materials and questions, as well as extensions and follow-up questions. Several sources for lab activities are listed in Appendix B and many of the activities found in NCTM's *Student Math Notes* and *The Mathematics Teacher* provide lab-type activities.

FIGURE 3.3. LESSON PLAN INTEGRATING LAB ACTIVITIES

Type of Class Activity	Minutes
Warm-Up	5–10
Lab Activity	80–90

- Introduction: Why?
- Data Collection
- Data Analysis
- Conclusion
- Closure/Wrap-Up
- Pulling the Lesson Together
- Extension

Total Time	**85–100**

ONE TEST-DAY SCHEDULE

The last time management plan may be used for test days. Figure 3.4 shows how time can be allotted so that the entire block is not consumed by test taking and an entire class is not spent reviewing for the test. In an alternate-day model, this plan allows for pulling the unit together. During the first part of the class, students work together to review the material for the test. Planning a structured activity for students either in the form of a set of review problems they work together as a partner aspect of the test, or in some type of game format, prepares students for the test. Teachers who use manipulatives and labs for data collection and analysis and/or who use technology can incorporate these methods of assessment into their testing procedures. Students then learn to value all aspects of the class. This type of plan makes the review student-centered and the structure focuses students. Also, if they know that there will be a structured review and are assigned a role or task before the class, most will not wait to cram immediately before the test and the time will be used wisely.

FIGURE 3.4. ALLOCATION OF TEST-RELATED TIME

Type of Class Activity	Minutes
Test Review	15–40
Test	60–85
Total Time	**85–100**

In a 4/4 plan or 75-75-30 plan, because teachers see the students everyday, the test might logically fall at the beginning of the class, using the second half for new material. Because time is so precious, an entire class generally should not be used for testing. The new unit could be introduced using an appropriate exploratory lesson at the conclusion of the test. Students who do not finish within an allotted time need to make arrangements to finish later.

Time management plans should be adjustable, not cast in stone. Those presented here are models of possibilities for arranging time during a block. As teachers, we encourage you to remain flexible and "try on" different time plans to see which ones work best and under what circumstances.

MANAGEMENT STRATEGIES

"I felt like I was a first year teacher all over again," many teachers reported after switching to a block schedule. Why? It was because of the great amount of time they devoted to planning and rearranging the curriculum for teaching in a block. How long did the feeling last? It took about six weeks for teachers to make the adjustment. Students seemed to have an easier time and adjusted in less time.

So what are some of the management strategies teachers have used to help them? Many work not because of the block but because they are good strategies. Many developed out of a need to control changes, but made teaching and planning more effective and efficient.

OLD LESSON PLANS

Many of the lessons and teaching strategies used before block are sound and should not be abandoned. Teachers should reevaluate them and incorporate them into their new block lesson plans. Figure 3.5 is a traditional lesson plan designed for a middle school prealgebra class that is investigating prime and composite numbers.

With a few changes, this lesson can be refurbished and used almost intact. It can be extended to review or introduce the concepts of least common multiple (LCM) and greatest common factor (GCF). In *How to use Cooperative Learning in the Mathematics Class* (Artzt and Newman, pp. 42–45; available from NCTM), there is a cooperative learning activity complete with teacher and student instructions plus a student data collection sheet ready for reproducing. We have used this activity to expand this lesson for a prealgebra and an Algebra I curriculum.

Teaching in the block should not be an overwhelming enterprise in which teachers throw out all of the previous "tried and true" lessons and create all new ones. It should encourage teachers to remodel the lessons that were good, incorporating some new activities into them. Working together with other teachers who teach the same courses helps.

FIGURE 3.5. TRADITIONAL LESSON PLAN FOR PREALGEBRA CLASS

Lesson Title (Optional) and Topic: Sieves and Trees

Lesson Objective(s): to find the prime numbers between 1 and 100

Materials Needed: a hundred's table (Sieve of Eratosthenes) and colored pencils

Prerequisite Skills/Knowledge: prime number, composite number, factors, multiples

Instructional Strategies: • Direct Instruction • Whole and Small Group Interaction

Warm-Up Activity: 5–7 fractions to simplify, some are already in lowest form

Material for Guided Practice: extend the list of primes through 200

Homework: find the prime factors of the composite numbers through 50

Lesson Extension of Activity: investigate ways to find prime numbers... Mersenne.

Assessment of Objective(s): find the prime numbers less than 50

CREATING FRIENDLY FOLDERS

One management strategy that aids both preservice and veteran teachers employs file folders to order and organize individual lessons. Teachers create a single folder for each lesson design with a complete lesson plan, including handouts and other activities inside the folder. The title of the lesson is written on the folder tab. At the conclusion of teaching each lesson, they make notes about what worked well, what did not work, and any inspirations that occurred during the lesson. They write them down and include them in the folder. The lesson is ready to use again immediately or in the future and it is accessible for sharing with colleagues.

Teachers lose their thoughts and ideas by waiting until the next year to pull the lesson out of the file cabinet. Recording ideas for improving the plan or adding new ideas or noting special considerations should be done as close to the completion of the lesson as possible so that the information is beneficial to teacher and students.

HOMEWORK

Teachers, students, and parents express concern about the implications which longer blocks of class will have on homework assignments. As each lesson is different, each homework assignment will be different.

Doubling the amount of time students spend in a class does not mean that the amount of homework should be doubled. Many teachers use class time to incorporate much of the "drill" that traditional homework includes. Out-of-class assignments then become more application-oriented, or extensions, which include the drill items for reinforcement.

There is no algorithm for assigning homework. Generally, instead of doubling the amount, many teachers find that they assign about one and one-half times as much per block. Depending upon the class, assignments include some skill-oriented problems and concentrate more on "word problems" and connections to other mathematical concepts and other disciplines.

When students come back to class after completing an assignment, it is not necessary to double the amount of time answering questions on homework problems. To do so wastes time that can be spent on more worthwhile tasks. Ten to 15 minutes is usually adequate. Unless the assignment is particularly arduous or students are very confused, minimal time should be spent checking homework because students become disinterested and lose their focus.

Several methods for handling homework review are:

- Have students list the problem number(s) they would like to discuss on the chalkboard. Each student puts a tally mark beside ones that are the same so the teacher can determine which problems created the most difficulty for students. The teacher then discusses those "most" requested. An alternate method from teacher-centered review is to have students volunteer to write their solutions, complete with reasoning on an overhead transparency sheet and then explain the problem to the class.
- Have students confer in groups to reach consensus and discuss the problems while the teacher circulates around the room to listen to discussion and offer assistance when needed.
- A list of answers can be written on an overhead transparency sheet and shown to the students. (We find this the least effective, because explanations are not written.)

ABSENCES

STUDENT ABSENCES

When students are absent, they miss an equivalent of two traditional classes. This can be problematic for students who have an extended absence or who exhibit chronic absenteeism because the amount of work to be made up mounts quickly. There are no easy solutions to this problem, but there are a few classroom management strategies that can be used to help students:

- Pairing students as study partners and having them exchange phone numbers affords them an opportunity to call each other if they miss a class.
- Having students take turns keeping a class log of notes and activities, complete with handouts, provides much needed information for the returning student.
- Posting a class assignment sheet allows all students the chance to keep their assignments current.

There is no substitute for having students in class, but getting assignments to them before they come back or getting them caught up as soon as they return can be made easier.

TEACHER ABSENCES

When a teacher needs to be absent, it is difficult enough to provide a lesson plan for a traditional class period, but a block class presents greater challenges, because the substitute cannot be counted on to know anything about mathematics and may be overwhelmed by the mere idea of being in a mathematics class. Some mathematics departments prepare "emergency" plans for the possibility of a teacher not being able to put a plan together. These tend to be generic in nature but do not have to be considered a waste of time. Using a video or series of videos with an assignment designed to go with them often works well. It provides an opportunity to make connections to other disciplines and possibly to incorporate a short-term project or writing activity. These types of lessons must be carefully planned well in advance with a copy of the video and other necessary materials readily accessible.

For absences that are known about in advance, a teacher can plan for the implementation of any lesson that is appropriate to the unit being studied. Working with a few students who can function as leaders of groups or of the class can alleviate many concerns a teacher may have about being away from the classroom. On these days, plans that incorporate group activities help move the class forward without wasting the day. If the directions for the substitute and the students are clearly written, the day the teacher is out might be used for data collection and analysis in a lab activity.

Lessons that do not work particularly well but are sometimes unavoidable are those that request students to "Read the following pages and then work the following problems." If the text material is not easily readable, students experience frustration and are not be able to work the exercises, so the day is wasted.

CONCLUSION

- Don't burn your candle at both ends! Work with other teachers who are teaching the same course; develop lessons, assessments, and activities together. Two heads are almost always better than one.
- Successful teaching in a block schedule is not about presenting two lessons one right after the other. Look for opportunities to weave concepts together in a meaningful way.
- Don't reinvent the wheel. Many previous lessons may just need a little cosmetic makeover with a few minor adjustments to give them "block vitality."
- "Variety is the spice of life" so strive for three activity changes during a class to keep the students active and the class alive. However, don't overdo it and leave your students wondering what they were supposed to learn. Be sure the changes make sense and that transitions are made smoothly. Changing activities just for the sake of changing activities makes no sense. A few well-planned activities work better than a multitude without a focused target.
- Include plans for your students to provide evidence of the concepts they have learned. Plan to assess students in the same ways they received instruction. If labs and calculators are integral to concept building, they should be included in a plan for assessing learning.
- Incorporate ways to check for understanding during each class. Moving around the room to watch student interaction and work gives an instant informal assessment. A short quiz or writing assignment may help both the teacher and the students to clarify understanding.
- Laboratory activities are an excellent method for incorporating many skills and concepts into the curriculum. Try a few simple ones and work with other teachers until you feel comfortable. Ask a science teacher for help. Include prelab activities to make the lab's purpose clear and postlab activities to pull all the concepts together in the end.

4

INSTRUCTIONAL STRATEGIES USEFUL FOR BLOCK LESSON DESIGNS

WHY VARY INSTRUCTION?

With longer chunks of time devoted to each subject, block schedules can be a catalyst for classroom innovation. The longer class period provides opportunities for innovative teachers to blossom and gives those teachers who traditionally use the a direct instructional style encouragement to include a variety of teaching methods. Block scheduling provides alternative blocks of time but it is what happens in each classroom that really counts. As stated in the *Curriculum and Evaluation Standards for School Mathematics* (NCTM, 1989, p. 125),

> A variety of instructional methods should be used in classrooms in order to cultivate students' abilities to investigate, to make sense of, and to construct meanings from new situations; to make and provide arguments for conjectures; and to use a flexible set of strategies to solve problems from both within and outside mathematics. These alternative methods of instruction will require the teacher's role to shift from dispensing information to facilitating learning, from that of director to that of catalyst and coach.

We believe that block scheduling demands that we plan lessons to include a variety of instructional strategies based on concepts rather than on one textbook section at a time. This chapter provides instructional strategies useful for block lesson designs.

Critical to the success of block teaching is varying teaching strategies within the period. Changing activities several times during a block prevents student and teacher boredom. Strive for three changes to keep students' attention and focus. It is a way to address different learning styles, to promote greater mathematical understanding, and to incorporate real-world applications and mathematical connections. It also allows the teacher an opportunity to build reinforcement activities.

This chapter presents instructional strategies. Within the discussion of each strategy are examples of math subject areas from middle school math to algebra, geometry and trigonometry through calculus. Individual lessons with incorporated strategies are provided in Appendix A.

USING WARM-UPS

Beginning each class with some type of warm-up activity sets the tone for the day's class. As students enter our mathematics classrooms we want them focused and engaged in the business of mathematics right away. We want activities that get them involved and thinking, get them "warmed-up." Starting as soon as they walk in the door helps them see that you view the time you have with them as valuable and important.

Warm-up activities are multipurpose and multidimensional. They include journal writing, SAT preparation questions, homework quizzes, puzzles, short quizzes, nontraditional problems, or a summary of a procedure or algorithm. They should vary according to the type of class taught and the needs of the students at the time. It may be a good time to review past concepts needed for an upcoming unit or to find out if the students understood their homework assignment, as a means to revisit concepts from previous lessons, or to lead into a new concept.

The time allowed for warm-ups usually ranges from 5 to 15 minutes at the beginning of the class period. During this time the teacher can take care of administrative duties, check roll, do a homework check and note student difficulties, make one-on-one contact with a student who was absent by giving instructions on missed work, and pass out graded papers. These few minutes have students on task and allow the teacher more instruction time during the period.

The following examples illustrate a variety of warm-up activities and the context in which the technique is used. These provide concrete examples and should inspire you to take ownership of them and to try some of your own.

DAILY QUIZZETTE

While teaching a group of ninth grade students in Algebra I, Part 1, we found that the students needed repetition on many basic skills. The students had a wide range of ability and basic skills. For a warm-up activity we designed a short quiz of five questions covering five topics. For four class periods we had their quiz sheet on their desks when they arrived in the room. Students had five minutes to complete the quiz and to turn it in. The next day the quiz sheet with the graded quiz was on their desks along with the new quiz on the same five topics. This type of quiz was helpful in several ways: students could see what type of questions were missed and knew immediately areas where help or practice was needed; also, we could see what type of help each student needed and we could identify areas of strengths and weaknesses; this information helped assign students to groups and study pairs.

If a student is absent, the quizzette missed is not made up. Instead, we count the next one twice. Also, many students show a great deal of improvement during the block of quizzes for the two-week period in an alternate block schedule or in a week on a 4/4 schedule. To accommodate for this improvement, we periodically allow them to count the last quiz for the entire block of quizzettes. Students are rewarded for improving and students that have been doing well all along are not penalized.

Figure 4.1 (on the next page) is an example of one of the teacher master quiz sheets to make into a transparency and Figure 4.2 is an example of a master student answer sheet to copy for each individual student.

FIGURE 4.2. STUDENT ANSWER SHEET

Name	Period
1. 2. 3. 4. 5.	1. 2. 3. 4. 5.
1. 2. 3. 4. 5.	1. 2. 3. 4. 5.

FIGURE 4.1. TEACHER'S MASTER QUIZ SHEET

Quizzette	Week of ____________
1. Solve for x: $2x + 4 = 12$ 2. Simplify: $6x + 7y + 13x$ 3. Evaluate: $9x - 7$, if $x = -3$ 4. Subtract: $-22 - (-51)$ 5. Add: $(2x + 5) + (-3x + 8)$	1. Solve for x: $3x + 4 = 10$ 2. Simplify: $16x + y + x$ 3. Evaluate: $4x + 7$, if $x = -3$ 4. Subtract: $2 - (-91)$ 5. Add: $(2x - 15) + (3x + 8)$
1. Solve for x: $3x - 4 = 17$ 2. Simplify: $x + 3y + 5x$ 3. Evaluate: $6x + 8$, if $x = -4$ 4. Subtract: $60 - (-90)$ 5. Add: $(2x + 15) + (9x + 7)$	1. Solve for x: $3x + 4 = 16$ 2. Simplify: $6x + 33x + 5x$ 3. Evaluate: $5x + 18$, if $x = 3$ 4. Subtract: $-90 - (-80)$ 5. Add: $(5x + 25) + (7x + 7)$

JOURNAL WRITING

Individual student journals that are kept in the classroom are convenient for student retrieval and provide another warm-up activity opportunity. This activity normally asks for a response to a question or statement. As students enter the room, ask them to pick up their journals and prepare to make an entry. Begin by asking a question that requires the students to respond by explaining a process, concept, or activity. This activity provides the teacher with both individual and group feedback. It may be used to determine if an individual has the necessary understanding of the concept, skill, or procedure, as well as to answer the question, "Is the class with me?"

- Traditional Problems
  - After Prealgebra students have had some practice in solving equations in one variable, give them a typical problem from their assignment.

    Solve the equation $4x + 3 = -12$ *for* x

    Ask the students to write an explanation of the procedure used to solve this equation for x. This can also be done during other parts of the lesson to solidify their knowledge of an algorithm.
  - As students in Geometry are learning to do proofs, provide them with a statement to prove. Ask them to analyze the situation and devise a plan to prove the statement, and then to write the plan.
  - In Trigonometry, upon completion of investigating the effects of A, B, C, and D on an equation in the form $y = A \sin B(x - C) + D$, ask the students to write descriptions of these effects and to illustrate them with diagrams.
- Nontraditional problems

  The *Joy of Mathematics* and *More Joy of Mathematics* by Theoni Pappas are resources to which every math teacher should have access. They are full of intriguing problems, ideas, concepts, puzzles, historical notes, and games that pique the students' curiosity and encourage them to think in creative ways. The following are examples of nontraditional problems.
  - Topic: Order of Operations

    Problem: Use the numbers 1, 9, 9, 8 in order with any mathematical operation to write the numbers from 1 to 100. (This can be assigned to groups and/or changed to writing the numbers from 1 to 10 to accommodate the particular class or the amount of time the teacher wants to allow for the warm-up.)
  - Topic: Area

    Problem: If the price is the same, which is a better buy: a 10-inch round pizza or a 9-inch square pizza?
  - Topic: Volume or Proportion

    Problem: If a jar of peanut butter that is 3 inches in diameter and 4 inches high sells for 60 cents, what would be a fair price for a jar that is 6 inches in diameter and 6 inches high? (NCTM, *Mathematics Teacher* calendar, September 1983)

- Topic: Guess and Check or Problem Solving

  Problem: Once upon a time, there was a horse trader named C. Biscuit. He went around trading his horses until he had none left. He went to the first buyer and sold half of his horses plus half a horse; then to a second buyer and sold half of the remaining horses plus half a horse; he went to a third and fourth buyer completing the same deal as before. After the fourth buyer, he had no horses left. He was not a cruel man and never harmed a horse.Can you figure out how many horses he traded? (This was given us by a student and the problem's author and origin are unknown.)

Other sources for nontraditional problems are the *Mathematics Teacher*'s monthly calendar, *Middle School Mathematics Teacher*'s menu, and *World's Most Baffling Puzzles* by Charles Barry Townsend. Listed in Appendix B are numerous puzzle books and contest problem books that also provide excellent problems.

MOVEMENT

With longer blocks of time students and teachers welcome opportunities to get out of their seats and move around. Try to plan lessons to include activities allowing students to move about and change their seats. This can be done when students are asked to work in small groups or with a partner on a task then asked to join a larger group for a continuation of the activity. Students may present problems, solutions, or findings by going to the board or by using the overhead projector. Lab activities often require students to move about to collect data or to perform an experiment.

AN ACTION CHALKBOARD GAME

An example of an activity that is good for skill and drill is in the form of a game played at the chalkboard. The teacher prepares problems, definitions, or visuals and then puts all the answers on the board in random order. Students are divided into teams and the teams are lined up in order so that each student will have a turn to answer a question. The teacher reads the question and one student from each team races to the board and places his or her hand (sometimes we use flyswatters to make things interesting) on the correct answer. The first team member to swat the correct answer is awarded a point for his or her team. The team with the most points wins! This idea came from an English teacher who often uses it to review for tests. It can be adapted and used in almost any class or at any level. It certainly provides for movement and generates enthusiasm in the classroom as well.

TAKING STUDENTS OUTDOORS

Students enjoy a change of scenery—going outdoors and doing math is a real treat, especially when the weather is nice. In the spring, our geometry classes do a unit on trigonometry. After learning the right triangle definitions for sine, cosine, and tangent functions, we make clinometers (an angle measuring device) and use it to help calculate the heights of teacher-selected tall objects such as the flag pole, lights on the football field, the stadium seats on the top row, tall trees, and the gymnasium roof. (An adaptation of this activity can be found in *Discovering Geometry,* Key Curriculum Press, on pages 542–543.) Students working in groups of three or four use a measuring tape, the clinometer, and their knowledge of right triangles to gather the data outside then return to the classroom and use their data to find the requested heights of all objects. Each group is given a sectioned sheet to fill in with its data. Each student is required to make a drawing, label all known measurements, label unknown quantities using variables, set up the equation, show the solution, and then state the final answer.

This type of activity could be done using similar triangles to find the unknown height of a tall object. The height of the student and the length of the student's shadow form two sides of one right triangle, and the unknown height of the tall object and the length of it's shadow form the second right triangle.

Movement activities may include data collection activities in which students bounce balls, swing pendulums, do the wave, model a curve using the motion detector, or go to the football stadium to do human graphs, a circle walk, or math aerobics. Use your creative juices and those of your colleagues to find other methods of instruction and learning utilizing movement.

CONNECTING MATHEMATICS TO OTHER DISCIPLINES

Have you ever had one of your students ask, "When are we ever going to use this?" Of course, the student is referring to the particular concept that the class is learning at that moment. The more relevant and more useful something is to each of us, the more value we place on it. Students know that it is important for them to know how to read to survive in our society, but often they don't see the real importance or relevance in learning mathematics nor do they see that it is more than learning a bunch of number facts. It is important to engage students in tasks that make mathematical connections to other disciplines. Connections should occur often enough for students to value mathematics and its relevance to real-world situations. Students need to view mathematics as something that has real value and enables them to apply it to a variety of problem solving situations.

This notion of connections is one of the Standards for School Mathematics. In *Curriculum and Evaluation Standards for School Mathematics* (p. 148), *Standard 4: Mathematical Connections* states:

> The mathematics curriculum should include investigation of the connections and interplay among various mathematical topics and their applications so that all students can—use and value the connections among mathematical topics; use and value the connections between mathematics and other disciplines.

Standard 4 goes on to say:

> Students' understanding of the connections among mathematical ideas facilitates their ability to formulate and deductively verify conjectures across topics, an activity that becomes increasingly important in grades 9–12. In turn, these newly developed mathematical concepts and procedures can be applied to solve other problems arising from within mathematics and from other disciplines.

It is up to us as teachers to seize every opportunity to help students make those connections and provide learning environments to stimulate students and simulate real-life problems that engage students in exploration, discovery, and extensions. The following is a brief list of applications making connections between mathematics and other disciplines. The applications are from *Curriculum and Evaluation Standards/Grades 9–12* (NCTM, 1989, p. 147):

- Art: the use of symmetry, perspective, spatial representations, and patterns (including fractals) to create original artistic works (NCTM 1989, p. 147) such as: the use of the golden rectangle and the golden ratio in Salvador Dali's "The Last Supper," which is displayed at the National Gallery of Art, Washington, DC; the development of linear perspective in the Renaissance using the movie *Masters of Illusion* from the National Gallery of Art; the use of modular arithmetic, Algebraic properties and symmetry to produce original works of art; the use of knowledge of polygons to make original tessellation design to put on fabric such as T-shirts or to make stained glass window decorations.
- Biology: the use of scaling to identify factors that limit the growth of various organisms; "Why King Kong couldn't exist," from *For All Practical Purposes* (COMAP); comparing a Lilliputian to Gulliver from *Gulliver's Travels* by Jonathan Swift, *For All Practical Purposes*

(COMAP); to use a Punnett square to show possible combinations of genes for one or more particular genetic traits.

- Business: the optimization of a communication network; use optimization techniques to solve business problems such as minimizing cost or maximizing profit under pre-established conditions.
- Industrial Arts: the use of mathematics-based computer-aided design (CAD) in producing scale drawings or models of three-dimensional objects such as houses.
- Medicine: modeling an inoculation plan to eliminate an infectious disease; using data on the spread of a disease to predict at what rate the disease is spreading.
- Physics: the use of vectors to address problems involving forces; gathering data from the CBL using the motion detector to find the location function of a particle moving in rectilinear motion; simple harmonic motion using a pendulum.
- Social Science: the use of statistical techniques in predicting and analyzing election results or population growth; interpreting graphs used in election speeches and debates; looking at a period of time in history and finding what inventions were made and what impact they have had on society.
- Science: to study the geometric shapes found in nature such as shells, spider webs, flowers, chambered nautilus, starfish, pine cones, honey combs, soap bubbles, gem stones, and pineapples; to study rates of change such as velocity and acceleration in Physics; to explore the graphs of sound waves with the help of Trigonometry.
- Literature: generate problems using the short story "The Pit and the Pendulum" by Edgar Allan Poe (refer to *Mathematics Teacher,* November, 1975); the short story "The Gold Bug" by Edgar Allan Poe, in which a message is decoded based on the frequency of use of letters in the English alphabet; detective stories (refer to *Mathematics Teacher,* November, 1975); and the poem "The Chambered Nautilus" by Oliver Wendell Holmes connecting the spiral in the nautilus and logarithmic functions. Other examples of literature that provide an abundance of sources for problems are *Through the Looking Glass* by Lewis Carroll; *A Tangled Tale and Other Knots* by Lewis Carroll; "Henry Wadsworth Longfellow, Poet Extraordinaire," by Charles Mitchell in *The Mathematics Teacher,* May, 1984, pp. 378–79; and Sir Arthur Conan

Doyle and the Sherlock Holmes stories to explore logic in *Discovering Geometry* by Michael Serra, 1993, pp. 565–71.

In order to provide examples that connect a teaching discipline other than mathematics to a particular math subject, we include examples relating the two disciplines under the math subject area heading.

MIDDLE SCHOOL

- Art: Use the properties of closure and mod arithmetic to make geometric designs resembling quilt designs. These designs could be further developed by repeating the design or using different types of symmetry and rotations. Both the art teacher and the math teacher working together could develop this into an interdisciplinary unit.
- History (Social Studies): Different types of information are displayed in the form of graphs and charts in history and social studies. By collecting data and making graphs to represent the relationship between the two sets of data, the math teacher can help students understand how to read graphs, how to interpret the information the graphs reflect, and how to make predictions based upon the information. The data collected could be populations of different countries, amounts of products that are imported or exported to and from different countries, wages of workers in different parts of the world, and/or numbers of people in different religions. By consulting the social studies teacher in advance, relevant data could be used to enhance the social studies lessons using mathematics.
- Science: Many schools have science fairs or have students do science projects without competing. A joint venture between the math teachers and the science teachers to work toward a common goal, that of teaching the students the scientific method and using it as a basis for their projects, could help the students focus on the gathering of data, observing the data, making conjectures, and arriving at a conclusion. Working together in the computer lab the teachers could show students how to make spreadsheets, graphs, and charts to display the data and the relationships found.

ALGEBRA I

- Biology: Scale and proportion can be used to compare the brains of different animals, for example, and to contrast the intelligence of one to another. Also, when students study genetics, show them how to set up the Punnett square so they can see the possible combinations of genes in one or more particular traits, such as blue eyes and brown eyes.

- Forensic Medicine: The news and many fictional television shows are filled with acts of crime and how they are solved. The shows often include references to the coroner or forensic scientist and how they go about solving a crime and establishing the time of death. When a skeleton is found, a forensic scientist uses bone calculations and formations to help identify victims of crime, the cause of death, and to help establish how the crime may have been committed. For an example of mathematics in forensic science refer to *The Applications of Secondary School Mathematics*, (NCTM, 1991, pp. 31-32). Another topic that uses mathematical relationships is how fast the body decays at different temperatures. How does a coroner estimate the time of death when it is not otherwise known?
- Health: Students are being bombarded with information about fat, calories, exercise, and staying fit. Problems that involve equations of several variables often have as their premise a need to find an appropriate balance in the values of the variables. One such problem would consider the nutritional content of the foods that we eat and compare that to what is recommended for the age group of our students. Refer to *The Standards, Algebra in a Technological World*, (NCTM, 1996, pp. 87–90).

GEOMETRY

- Art: Geometry is a natural vehicle for studying art or for enhancing the study of geometry through art. Salvador Dali's "The Last Supper" is a study of mathematics in art. He utilized the meaning of shapes as well as the shapes themselves. The painting has line symmetry, linear perspective, symbolism, the golden ratio and the golden rectangle. The art of making tessellations, as done by M.C. Escher, uses all types of regular polygons. Students could make their own tessellations and put them on T-shirts or make stained glass windows out of 8-inch by 10-inch glass taken from inexpensive picture frames. The glass designs could be placed in the frames for display in the classroom and then taken home.
- History/English: In our school, ninth graders in the honors program take World History and English as a course designed to integrate the two. Many of these same students take Honors Geometry. As the

students study the Renaissance period including the art of the Renaissance, they could also be introduced to the mathematics and the mathematicians of the time. Renaissance art and architecture made use of the golden ratio and the golden rectangle and some of the most phenomenal mathematics was revealed during this time. Working with teachers of these two subjects would make an ideal interdisciplinary unit.

- Biology: Geometric shapes are found everywhere in nature. These shapes are found in starfish, sea shells such as that of the chambered nautilus, pine cones, bee hives, flowers, leaves, soap bubbles, crystals, gem stones, and pineapples. Cut an apple in half horizontally instead of vertically and you find a star or pentagram.
- English: As students are learning to write proofs in geometry and wondering why it is important to do proofs, a natural connection to be made is that between writing a informative essay and writing the geometric proof. By comparing the two, the student is able to make this connection. The thesis statement in the essay relates to the statement of what is given and what is to be proven in the formal geometric proof. The role of the thesis statement is to serve as a connecting thread that unifies the essay. It contains the topic, a focus, and a purpose. The given and the prove statements do the same in a proof. Each following paragraph contains examples, reasons, facts, and details directly related to the topic and in turn to the thesis statement. The body of a proof consists of statements and reasons that support the given and direction of the proof. These also have to be in logical order for them to be coherent. The final paragraph in the essay is the conclusion, which serves to bring the discussion to a logical end. It reinforces the writer's ideas. The conclusion of a proof brings the discussion to an end by making a "call for awareness" and making the point of the proof the focus of the conclusion. *Simon and Schuster Handbook for Writers*, (Lynn Quitman Troyka, 1990).

ALGEBRA II

- Biology: By using scale, students can identify factors that limit the growth of various organisms. "Why King Kong couldn't exist" from *For All Practical Purposes* (COMAP), is a lesson designed to integrate biology and mathematics. Comparing a Lilliputian to Gulliver from *Gulliver's Travels* by Jonathan Swift by using scale. helps in understanding the same limits.

- Literature: A source for algebra and geometry problems can be found in *Pillow Problems and A Tangled Tale* (Lewis Carroll, Dover Publications Inc., 1958).
- Science/Astronomy: As students study conic sections, they develop an understanding of the relationship between the shapes of the conic sections and the shapes found in the mirrors in telescopes, in speakers hooked to stereos, in the headlights on automobiles, in optics, in photographic equipment, and in the orbits of the planets.

TRIGONOMETRY

- Physics and Science: The trigonometric functions form the basis for studying many of the topics in physics and science such as periodic real-world phenomena, harmonic motion, sound waves, uniform circular motion, temperature changes, biorhythms, and tide variations.

PROBABILITY AND STATISTICS

- Social Sciences: Can the public place any confidence in the results of polls such as the Gallup Poll? How are these polls done? As students learn how statistics are developed, these types of questions can be answered. Students are able to see how predictions are made based on the polls. They can analyze newspaper articles that present statistical data.
- History: Have students research a period of time in history and find what inventions were made and what impact they had on society. For example, research the time of Sir Isaac Newton to see what advances were made in mathematics and what impact these revelations had on society of that time and on future generations.
- Marketing: What is the cost of advertising during the Super Bowl? If you owned a chain of food stores how would you determine what products would sell best at each individual store? These questions require the student to think about all of the factors that would effect the outcome, then require them to think about how they can address each issue, and then require them to find the process that would help them find the information needed. Would it help to survey the area where food stores are located to find out the kinds of foods that are popular? Are there any cultural events or traditions that require certain kinds of food at particular times of the year? Does it help to advertise and, if so, what impact does it have on the cost of a single

item? Students could generate surveys and then collect data, analyze it, and draw conclusions.

PRECALCULUS AND/OR CALCULUS

- Physics: Use the Calculator-Based Laboratory (CBL) to investigate sound waves, to analyze a falling object, to investigate Newton's Law of Cooling, to analyze fluids, force, and pressure, and to develop the concepts of velocity and acceleration by using a motion detector.
- Business: Optimization problems concerning profit, cost, revenue, and the factors that effect them are found in most Calculus texts.
- Biology: How do we predict how fast an epidemic will spread? How do we estimate populations of people and animals and make predictions about the rate of population growth?
- History: Discuss Newton's problem that started Calculus and then discuss its impact on society.

CONCLUSION

Part of the vision of the National Council of Teachers of Mathematics is "a curriculum for all that includes a broad range of content, a variety of contexts, and deliberate connections." By making connections we are helping students value mathematics and its relevance to real-world situations; and we are preparing them for the type of problem-solving needed for them to be contributors in our society.

DISCOVERY, INQUIRY, AND INVESTIGATIONS

Discovery is defined as the process of obtaining sight or knowledge for the first time or the process of making known or visible. It requires exploration and investigation while forcing the student to think, to draw conclusions, and to make generalizations. One of its main advantages is that it motivates students toward learning broad concepts and principles and encourages them to communicate conjectures or conclusions to others.

Discovery, inquiry, and investigative learning activities involve students in the process of discovery by enabling them to collect data, observe patterns, test hypotheses, and make conjectures or draw conclusions. These strategies are student-centered.

Thinking back to our own experiences in math classes, mathematics itself was a discovery to us. As we found patterns in numbers, or worked "story problems," or wrote a formal two-column proof, we felt as if we had discovered each one all by ourselves. Even though others had made the discoveries before us, we did not know that they had nor did we know how they had done it. We felt ownership. This is one of the joys in learning mathematics. We want our students to find this joy of discovery. Providing activities and opportunities that allow the student to use his or her present knowledge and ability to collect data and analyze it is a strategy to use often.

"Bruner suggests that students learn through their own active involvement with concepts and principles and that they should be encouraged to have experiences and to conduct experiments that permit them to discover principles for themselves" (*Methods for Effective Teaching,* Bruner & Byrd). Students' curiosity is aroused and this motivates them to continue to pursue the process of finding relationships and even answers. It teaches problem-solving skills and encourages students to analyze data and to manipulate information rather than just absorb it.

Teaching geometric concepts provides the ideal situation for using the discovery method of instruction. Many text books incorporate the discovery method into student lab workbooks, enrichment activities, and investigations. Two texts that provide numerous discovery activities and lessons are *Geometry* (Addison-Wesley) and *Discovering Geometry* (Key Curriculum Press).

The Addison-Wesley text has a lab workbook and an enrichment workbook, as well as technology activities, practice worksheets, lesson plans, and evaluation materials. (These lessons are designed for use in the traditional 50-minute class period; however, they are adaptable to a longer block of time.)

The *Discovering Geometry* text is exactly what the title implies. It is full of investigations that lead to geometric discoveries within each lesson or unit. It also contains cooperative learning group activities at the end of each unit that are designed to engage students. Along with the *Discovering Geometry,* Key Curriculum also publishes *Patty Paper Geometry* by Michael Serra. It contains construction investigations using patty paper that lead to geometric discoveries.

These activities can be done either by individual students or in groups. If done individually, students probably will need more time and guidance from the teacher. According to Michael Serra,

> Cooperative learning is the most effective mode of classroom instruction for doing investigations in *Patty Paper Geometry.* The best group structure for patty paper geometry is pair-share. In pair-share, students who ordinarily work in groups of four break into two pairs. One student in each pair reads the instructions to the partner while the partner does the folding. The pair compares its results with the results of the other pair in the group. The group then makes its conjecture. For the next investigation, both the pairs and the roles switch.

> This cooperative group structure will help you get all your students involved in sharing the excitement of discovery learning. In pair-share, everyone has a role and shares in the pride of discovery. (*Patty Paper Geometry*)

Effective teachers need a repertoire of instructional strategies to use with students because students come from varying educational backgrounds and have different learning styles. In a block schedule, it is important for teachers to vary the instruction to maintain momentum, to accommodate different learning styles, and to keep the student's interest. The discovery, inquiry, and investigation method provides activities that enable the students to do mathematics andto have a better chance of understanding it .

The following are activities that use discovery, inquiry, and investigation.

GEOMETRIC CONCEPTS (MIDDLE SCHOOL OR HIGH SCHOOL)

- Concept: Angle relationships formed when two parallel lines are cut by a third line (a transversal).

  Activity: Prepare a discovery sheet with a diagram of two parallel lines already drawn and cut by a third line. Number all of the angles. Make a chart with one column containing blank spaces for the measure of each angle, another column pairing the angles that have the same measure, and a third column for those that have the sum of the measures equal to 180 degrees. Students then draw their own diagram having two parallel lines cut by a third line and number their angles. Students are to make conjectures about which angles will always have the same measure and which angles will always have a sum of 180. Students then draw two lines that are not parallel with a third line intersecting these two lines. Are the angles in the corresponding locations to the other drawings also equal in measure? Follow up with a discussion of the conjecture about the angle measures and about the importance of the conditions necessary to make it true.

- Concept: The sum of the measures of the angles in a triangle is 180 degrees.

  Activity: Prepare a sheet of paper with several triangles of different sizes and shapes drawn on it. Be sure to name the triangles. Include a chart to be used by the students to record the measure of each angle for each triangle and also the sum of the three angles for each triangle. Ask the students to make a conjecture about the sum of the measures of the angles in each triangle. Discuss why all students did not get the same exact answer but everyone got very close to the same answer.

- Concept: The sum of the exterior angles of a convex polygon is 360 degrees.

  Activity: Prepare a sheet of paper with several convex polygons drawn on it, some regular and some irregular polygons. Starting at one vertex draw an exterior angle and continue to draw one exterior angle at each corner by extending each side of the polygon in one direction to form the angle. Students are to measure each exterior angle and then find the sum of these angles. Ask students to make a conjecture about the sum and then to write an explanation of how they came to that conjecture. The written explanation could be done each time students are asked to make a conjecture or, on occasion, to add variety and to encourage students to think through the processes of examining what they know and of arriving at conclusions based on what they know and by using logic.

- Concept: The side opposite the largest angle in a triangle is the longest side; the shortest side is opposite the smallest angle.

  Activity: Prepare a sheet of paper with several triangles of varying sizes and shapes drawn on it. Label the triangles. Include a chart for the students to fill in with the measure of each angle of a triangle and the length of each side of the same triangle. After students have finished finding the measurements for each triangle, have them make a conjecture about the length of a side of a triangle as it is related to the opposite angle measure compared to the other angle measures in the same triangle.

- Concept: If you know the lengths of the sides of a triangle but not the angle measurements, how do you determine if the triangle is acute, obtuse, or right?

  Activity: Prepare a sheet of paper with several triangles of different sizes and shapes drawn on it. Be sure to include triangles that are acute, obtuse, and right. The lengths of the sides may be stated or the students may be asked to measure each and record the results. Provide the students with a chart to record the results. On the chart be sure to include a column heading for the square of the length of each side as well as a place for the students to find the sum of the squares of the lengths of the two shortest sides. Students are to compare the type of triangle and the relationship of the sum of the squares of the two shortest sides with the square of the longest side. This activity is recommended for use in groups or with a partner. Students are to make a conjecture based on their findings.

For other geometric activities refer to *Patty Paper Geometry,* by Michael Serra (Key Curriculum Press, 1994).

TRIGONOMETRY

- Concept: The effects of *a, b, c,* and *d* on the graph of the equation in the form of $y = a \sin b(x - c) + d$.

  Activity: Using a graphing calculator have students graph several equations of the form $y = a \sin x$, changing only the value of *a* in each equation. Be sure to include fractions and both positive and negative numbers. Have students make a conjecture about what effect *a* has on the graph of $y = \sin x$. Repeat this again but this time use equations of the form $y = \sin bx$. Have students make a conjecture about the effect *b* has on the graph. Continue this process with the other two constants *c* and *d*. Follow with graphs of equations with more than just one letter replaced by a number. Ask students to predict what the graph will look like before they sketch it on the graphing calculator. This activity could also be done in student groups, each of which is assigned one of the equations with numbers to be placed into the position of one letter. Each group then could explain the effect that its assigned letter had on the graph of the equation. After all groups have finished an equation in the original form with all letters *a, b, c,* and *d* replaced by numbers, students could predict what the equation will look like.

ALGEBRA II

- Concept: Graphing parabolas-What is the effect of the value of *a* on the graph of the equation $y = a x^2$?

  Activity: Prepare a sheet of paper with several equations of the type $y = a x^2$ having different values of a and a large grid for the graphs. Be sure to include positive, negative, and fractional values of *a*. The students graph each equation on the graphing calculator and sketch the graph on the provided grid. By placing them on the same graph the students are able to compare one to the other. Ask the students to make a conjecture about the effect of a on the graph of the equation in the form $y = a x^2$.

Discovery, inquiry, and investigative learning activities can be the introduction to a new topic or a short activity combined with other instructional strategies. These activities also provide a method for instruction to involve the students in discovering for themselves, the principles which may take an entire block to develop. Because discovery encourages students to analyze data and manipulate information, it teaches problem solving. This method has a powerful impact on instruction and is more applicable in the block schedule than in the traditional class schedule.

CONCEPT DEVELOPMENT

Concept development, an instructional model developed by Hilda Taba, was designed to utilize the prior knowledge of students in developing, refining, and extending their understanding of concepts, both concrete and abstract. This model empowers students by letting them begin with a listing of "all that they currently know or think" about a subject, topic, or concept, and concludes with a synthesis and summary of the concept drawn from a refined list of ideas. We list the steps to the process in their entirety; but with practice and confidence in the value of this model, teachers may tailor it to meet the needs of both their students and the curriculum. For a detailed discussion of this model, see *Instruction: A Models Approach* by Gunter, Estes, and Schwab.

Before beginning the model, the teacher needs to explain the concept development process to the students, and let them know that all answers will be accepted. This relaxes students and makes all students partners in the process. Concept development begins with a whole-class activity, which leads to small-group work, followed by individual work, then back to a whole-class interaction, and which concludes with an individual assignment. While it is not necessary to include all of the steps in the model, it is important for teachers to learn the model pure without deviations so that it may be used appropriately for instruction.

- **Step One:** Ask students to silently brainstorm and then write down on a sheet of paper everything that they have ever heard, seen, read, known, or thought about a given topic or concept, for example, acid rain, which fits into a unit on logarithms in Algebra II or Precalculus. Give them two or three minutes to list their ideas while you walk around the room encouraging those who may be reticent or having difficulty. This allows every student to have something to contribute to the whole-group activity. Next have students raise their hands and offer **one** of their ideas, which you write on the chalkboard. Let the students know that there are no incorrect responses and all ideas are not only welcomed but encouraged. You should make no comments about the students' ideas with the exception of asking for clarification of any which seem unclear. "Encourage students to continue listing, even after they appear to have run out of information. Some items which follow the first pause are the less obvious ones, which frequently derive from greater insight and more thought." (Gunter et al., pp. 107–8) Before allowing a student to offer a second item from his or her list, try to write an idea from every other student in the class. Students will think of more things as the list grows if the topic is not too narrow or unfamiliar. The listing and recording should take less than 15 minutes. This is an excellent way to introduce applications into the classroom and to determine just how much students know about them.

- **Step Two:** After the list is completed, ask students to examine it to see if any of the ideas listed go together. Could they be grouped together under a common category. Ask them to identify those that go together. For the first group, write a 1 beside all that go in that group, then write a 2 beside those that fit in a second group, then 3 for a third group, and so on until all items have been placed into a grouping. Time should be allotted to discuss whether some items might belong in two or more groups or if some groups might be eliminated by subsuming them into other groups.
- **Step Three:** Put the students into groups of three or four, at least enough groups to cover all of the categories from Step Two. Ask groups to record all of the items from their category or categories on a sheet of poster paper with a marker and then find a label which fits the list. They should examine the list carefully and make sure that all items truly belong on their list and not another. When they reach consensus on a category label, they should write it at the top of their list. Each group tapes its list(s) to the wall or the chalkboard for the rest of the class to see. One representative from each group then explains how they came up with the labels.
- **Optional Reading:** (When teaching this model, Mary Alice Gunter added a reading component after Step Three which strengthened and reinforced the model.) Students are to read a teacher-prepared or other short article on the subject. As they read they should look for information that supports, refutes, or adds to the their generated list. Each student marks his or her copy appropriately to reflect his

or her findings. For example, a (√) could be used beside a passage that supports a listed item, an (x) could mark something that disagrees with a listed idea, and a (+) could signify new information. At the conclusion of the reading, ask students to contribute their findings by stating passages that confirm, contradict, or add information to their list. New information should be placed in the proper category.

- **Step Four:** At this stage, have students study and analyze the categories to determine if any categories need regrouping or inclusion within another category.
- **Step Five:** Finally, each student writes a brief response in which the information gathered and generated during the concept development process is summarized in work that synthesizes the separate parts into a whole.

"Concepts function like files in a storage cabinet—they provide a way of classifying and thus simplifying incoming information so that it may be stored in meaningful ways and retrieved later" (Gunter et al., p. 114). By manipulating information and reorganizing it into a new whole, students gain the advantage of constructing their own understanding.

Teachers use concept development for various purposes: to introduce a unit's primary concept and diagnose students' prior knowledge of that concept; to foster teamwork and encourage students to learn from one another; or to wrap up a unit in a culminating activity in which the teacher finds out what learning has occurred. Teachers should use this model throughout the year, not just one time.

Concepts or topics that lend themselves to the mathematics classroom include mathematics, statistics, acid rain, earthquakes, and patterns.

CONCEPT ATTAINMENT

The concept attainment instructional model enables students to "define concepts inductively." (Gunter et al., p. 90) In this model, students build a concept by carefully paying attention to the characteristics that set this concept apart from another concept. For example, a young child understands what is a dog and what is not a dog. Shown pictures, the child can distinguish a cat from a mouse or an elephant. Also, the child grasps the differences between the concepts of dog and cat if exposed to enough examples of both animals. The child can differentiate between "catness" and "dogness." Adults provide opportunities for children to correctly construct meaning for concepts.

In the classroom, mathematics teachers, with practice, can enhance the conceptual understanding of their students by providing structured environments in which students practice visualizing which attributes belong to a concept and which do not. Teachers must take care to select and write a proper definition of a concept, select its attributes, and develop positive and negative examples of the concept before entering the classroom. This model works best with those concepts that have well-defined sets of attributes. Despite the time it takes to develop an effective concept attainment lesson, the benefits to students are immeasurable. *Instruction: A Models Approach* carefully details this process.

- **Step One:** A teacher's first step is to choose an appropriate concept on which to build the attainment lesson. Geometry is rich in examples of concepts that have well-defined characteristics: symmetry, triangles, rectangles, and other geometric figures to name a few. Algebraic concepts include function, equation, polynomial, and sequence. Middle school attainment concepts could include prime number, composite number, pattern, and specific geometric figures. Once the teacher selects a concept that is teachable using this model, he or she must write a definition of the concept limiting the definition to the lesson being taught. For example, for the concept parallelogram, the teacher definition could be "any four-sided geometric figure whose opposite sides are parallel."
- **Step Two:** The teacher then uses the definition to pick out the essential attributes of the concept. For a parallelogram they are: geometric figure, four-sided, and opposite sides parallel. These characteristics set a parallelogram apart from any other geometric figure. If students can identify these, they own the concept.
- **Step Three:** Next the teacher prepares many examples of the concept for students to use in the process. They may be cut out rectangles, drawn rectangles, or pictures of rectangles as long as each of the examples contains *all* of the attributes. Also, the teacher prepares nonexamples (triangle, kite, dart, trapezoid, etc.) for students to use in distinguishing between what is and is not a parallelogram.
- **Step Four:** Now the teacher is ready to introduce the model to the class. Explain to the students that they are going to see positive and negative examples of the chosen concept. Their mission is to pick out the characteristics which separate the positive examples from the negative. Eventually they should be able to construct their own definition of the concept. On the chalkboard or wall, make two areas:

one for positive examples with space to write the positive characteristics and one for negative examples and their characteristics.

- **Step Five:** Put up a positive example and ask students to study it for a few minutes. Then ask students to describe the features of the example and record them in the positive area. Emphasize to them that all answers will be accepted so they will relax and contribute their ideas without fear. Record all responses. Later, some of the responses will be eliminated because they do not fit all of the examples or are not essential attributes. However, remind the students that every positive example contains all of the characteristics of the concept; whereas, the negative examples are missing at least one of the crucial features.

  More than likely, at least one student will yell out the concept. Record the response in the appropriate area but insist that the student(s) tell you what makes that concept what it is. It is also possible that the students may never arrive at the "name" of the concept even though they have discerned all of the essential attributes. This is merely a matter of the teacher providing the label for them.

- **Step Six:** The teacher puts the students into small groups and asks them to take the positive attributes and work together to write a definition of the concept. Students may become frustrated during this process because most of the time they are merely given a definition to memorize and to recall. Now they are being asked to create their own definition and form their own concept; not to simply take one from a dictionary, a glossary, or the teacher and accept it as truth. Encouragement is important and students should realize that the definition they write may not match the "traditional" one—it may be better!

- **Step Seven:** Once students write a working definition, the teacher gives them more positive and negative examples on which to test their definition. Students may be asked to create their own positive and negative examples.

- **Step Eight:** Just as in a laboratory experiment, bringing the students together to discuss what occurred during the process and how they arrived at their conclusion is vital. Ask each student to pinpoint the exact moment when he or she understood the concept being illustrated by writing about it in his or her notebook or journal. Providing opportunities for students not only to think but also to think about their own thinking is powerful in the learning process. Once they have constructed their own definition, they possess ownership of it.

WRITING AS A TEACHING STRATEGY

Writing is one way of making sense of what we have learned. It is a way to refine our thoughts and to make what we have learned our own. In the secondary classroom, we assume that our students will take notes of the key ideas, definitions, examples, and procedures to reinforce what they have learned for use as a reference, if needed, for clarification when working practice problems or for review. Writing is also an excellent way to communicate mathematics to others. It can be an important teaching and learning tool, as well as a method for providing variety in classes meeting in longer blocks of time. It can be used at the beginning of the class period as a warm up, during the middle of the block for reinforcement, or at the end of the block to summarize the important ideas for the entire block or unit.

Writing in mathematics is supported by NCTM's *Professional Standards for Teaching Mathematics* (1991) as an instructional technique because it helps students clarify meaning and communicate mathematics. This communication is vital to the process of concept development. It also provides opportunities for students to analyze and to use their reasoning skills. Writing is a form of student-centered communication in which students gain a better understanding of mathematical concepts.

In the *Professional Standards,* several of the standards address writing. Writing journals and original word problems are a way to respond to Standard 1, which suggests that the interests, understanding, experiences, and learning styles of students should be the basis for the selection of mathematical tasks for use in instruction, thus engaging students and building a framework for problem-solving in mathematics. (Neil, *Mathematics the Write Way, 1996. p.16*)

The *Professional Standards* suggests that teachers should pose questions and tasks that challenge student thinking and that ask students to clarify and justify their ideas in writing. It also suggests that the teacher promote classroom discourse using a variety of instructional tools. Writing is a tool that supports discourse. Other tools include using models, pictures, diagrams, tables, graphs, metaphors, analogies, stories and written hypotheses, explanations, and arguments in the mathematics classroom (NCTM, 1991, p.52).

Writing in the mathematics classroom also may be used as a way to assess individual student progress or the progress of the entire class. It is a means for the teacher to pinpoint problem areas that may need more clarification. We discuss this in more detail in the assessment chapter.

In *Mathematics the Write Way* (pp. 18–19), Neil makes the following general suggestions for writing in mathematics to both the middle school and high school teacher:

- Provide many opportunities for students to write in mathematics, not just during mathematics classes.
- Provide students with writing role models, such as those in good (children's) literature. Read to and with students, and have students read good literature.
- Provide students with many experiences for gathering information, such as hands-on activities, field trips, books, and magazines to give them backgrounds for writing.
- Provide many opportunities for interaction of students with other students and with the teacher for communicating and sharing ideas, concepts, and writings. These opportunities include working with partners, in small groups, and in whole group discussions.
- Give students many experiences with physical materials (including manipulative) to provide a background of experiences that promote communication among students and provide first-hand knowledge; both are necessary for writing.
- Have students write and illustrate their own story problems. Student-created stories have meaning for students.
- Have students brainstorm in small group or large group settings and record and classify terms or concepts to be used in their writings. Encourage students to discuss their classifications.
- Present many open-ended and/or probing problems for students to solve.
- Create a learning environment in which all students feel comfortable and where they can successfully do mathematics and write about their experiences.
- Integrate language arts with mathematics. Students can extend their writing skills in meaningful ways as they write in mathematics.
- Use the students' own language in developing mathematical problems for them to solve.
- Expose students to models of the kinds of problems they will write. Models should come from the teacher, other students, and from the formal curriculum.

Class writing is more than taking notes. Writing activities may include sentence completion, lead sentences, warm-ups, rewording, word banks, and debriefing. The following list, from Aggie Azzolino's *1990 NCTM Yearbook* article (p. 94), gives an abundance of ideas to explore in using writing in the mathematics curriculum.

- Out of Class
  - Writing term papers
  - Writing cheat sheets for tests
  - Completing statistics projects
  - Writing abstracts
  - Critiquing readings
  - Writing possible test questions
  - Making Journal entries—
    open-ended
    nongraded
    free format
  - Writing technical papers
  - Writing dictionaries
  - Summarizing readings
  - Completing take-home tests
  - Writing word problems
  - Writing letters to the teacher
  - Completing question books (Azzolino and Roth 1987)
    specific questions
    graded and regraded
    structured format
- In Class
  - Writing drafts of papers
  - Beginning class with warm-ups
  - Taking notes
  - Making journal entries

- Writing, exchanging, and critiquing word problems
- Doing short writings of sentences, phrases, lists, paragraphs
- Debriefings after an important point or procedure
- Debriefings at the end of the class

♦ On Tests

- Writing proofs stated in English or symbolically
- Writing or completing definitions
- Correcting false statements in true-false questions
- Explaining procedures
- Writing essays

One of the effective methods of using writing to help students form their own mathematical understanding is through journals. A journal is a diary-like series of writing assignments. It has two basic functions. First, it allows students to express their understanding in their own words. Second, it provides the teacher with a unique assessment tool. The entries usually take four to eight minutes to complete. The teacher may have students write in their journals every day to begin class or perhaps have them respond to the learning of a new concept at its completion.

While teaching Algebra I, we decided to incorporate writing as an instructional strategy in an effort to provide another opportunity for students to internalize mathematical concepts and to be able to communicate their understanding of those concepts. During this nine-week period we worked with a teaching associate from the University of Virginia. We were eager to try using journals with the students and working together, designed a format of questions and/or statements related to the unit that would be taught.

The method for evaluating the journals posed an important question. Realistically, we knew that to get thoughtful responses from the students we needed to read their journals and to use these writings to assess their mathematical understanding. We developed an assessment model for student responses for each unit we teach. We use a system based on three to five points for each student response.

We used journal writing as an in-class activity; however, it may be done as part of the homework assignment. Keith suggests using explorative assignments as part of a homework assignment ("Explorative Writing and Learning Mathematics," *Mathematics Teacher*, December, 1988). The assignments would include these types of exercises (Keith, p. 716):

- *Summaries.* Summaries can reveal where a class stands or serve as an introduction to a concept.
- *Visual image translation.* These experiences integrate visual and verbal understanding.
- *Synopsizing tactics for solving a problem.* These exercises enhance students' awareness of problem-solving techniques.
- *Giving an algorithm.* Students practice describing steps in algorithmic thinking.
- *Giving a definition or stating a theorem.* Such verbalizations enhance the precision of students' terms and theorems.
- *Communicating thoughts to a specific audience.* These exercises give students the experience of translating one explanation into another.
- *Inventing a problem.* Students learn to create their own illustrations through such exercises.
- *Generalizing a concept.* These exercises help to strengthen students' grasp of the overview of mathematics.
- *Group projects.* Such projects enhance confidence and productivity by having students work collaboratively. These group activities allow students to exchange ideas and to pursue the development of an idea in greater depth than when they are working individually. Because students enjoy this activity, it is a good way to introduce writing assignments.

For further examples of writing assignments and methods of assessing them, refer to the following: "Explorative Writing and Learning Mathematics," Sandra Z. Keith, *Mathematics Teacher,* December, 1988; "Using Writing to Learn Mathematics," Cynthia L. Nahrgang and Bruce T. Petersen, *Mathematics Teacher,* September, 1986; "Descriptive-Paragraph Miniproject," Cherlyn Kern, *Mathematics Teacher,* May, 1997; "Connecting Writing to the Mathematics Curriculum," David K. Pugalee, *Mathematics Teacher,* April, 1997; *Mathematics the Write Way,* Marilyn S. Neil, Eye On Education, Inc.; "Writing: A Necessary Tool for Learning," Wanda Leigh Elliott, *Mathematics Teacher,* February, 1996.

CONCLUSION

"Writing can be a viable force in the mathematics classroom both as a learning tool and as a means to assess a student's level of understanding" (Pugalee, *Mathematics Teacher,* April, 1997, p.310). For the teacher, writing as an instructional strategy offers variety, spontaneity, and a form of assessment of student learning. It provides opportunities for students to formalize and to communicate their understanding of mathematics. Writing should be encouraged in the mathematics classroom and be included as an important part of the mathematics curriculum.

COOPERATIVE LEARNING

Most people consider the learning of mathematics as a solitary process, one in which only a few talented people excel and the rest simply survive. There is little interaction and therefore students perceive math as something you do all by yourself.

Traditionally, many teachers rarely use group work or group activities in their mathematics classrooms. It is not the way they learned math nor was it a part of the way we were taught to teach math. Why should cooperative learning be a part of the way we teach mathematics? Because we are social beings and like interaction; we need to change these traditional perceptions. Small-group cooperative learning addresses these perceptions in several ways.

What is cooperative learning? According to Artzt and Newman, "Cooperative learning is an approach that involves a small group of learners working together as a team to solve a problem, complete a task, or accomplish a common goal. Group members must realize that they are part of a team and that the success or failure of the group will be shared by all members of the group. To accomplish the group's goal, students need to talk with one another about the problem and to help one another" ("Cooperative Learning," *Mathematics Teacher*, September, 1990, p.448).

Support for cooperative learning is stated in the NCTM's *Curriculum and Evaluation Standards for School Mathematics (Standards)*: "Small groups provide a forum in which students ask questions, discuss ideas, make mistakes, learn to listen to others' ideas, offer constructive criticism, and summarize their discoveries in writing" (p. 79).

In *Cooperative Learning in Mathematics*, (1990, p.4-5), Neil Davidson advocates the use of cooperative groups:

> 1. Small groups provide a social support mechanism for the learning of mathematics. Students have a chance to exchange ideas, to ask questions freely, to explain to one another, to clarify ideas and concepts, to help one another understand the ideas in a meaningful way, and to express feelings about their learning. This is part of the social dimension of learning mathematics.
>
> 2. Small-group learning offers opportunities for success for all students in mathematics. Students within groups are not competing one

against another to solve problems. The group interaction is designed to help all members learn the concepts and problem-solving strategies.

3. Unlike many other types of problems in life, school mathematics problems can actually be solved in reasonable lengths of time, such as a class period. Moreover, mathematics problems are ideally suited for group discussion in that they have solutions that can be objectively demonstrated. Students can persuade one another by the logic of their arguments.

4. Mathematics problems can often be solved by several different approaches. Students in groups can discuss the merits of different proposed solutions and perhaps learn several strategies for solving the same problem.

5. Students in groups can help one another master basic facts and necessary computational procedures.

6. The field of mathematics is filled with exiting and challenging ideas that merit discussion. One learns by talking, listening, explaining, and thinking with others, as well as by oneself.

7. The role of small groups in mathematical communication is addressed in the *Curriculum and Evaluation Standards for School Mathematics* by the National Council of Teachers of Mathematics (1989):

> Teachers foster communication in mathematics by asking questions or posing problem situations that actively engage students. Small-group work, large-group discussions, and presentation of individual and group reports—both written and oral—provide an environment in which students can practice and refine their growing ability to communicate mathematical thought processes and strategies. Small groups provide a forum for asking questions, discussing ideas, making mistakes, learning to listen to others' ideas, offering constructive criticism, and summarizing discoveries in writing. Whole class discussions enable students to pool and evaluate ideas; they provide opportunities for recording data, sharing solution strategies, summarizing collected data, inventing notations, hypothesizing, and constructing simple arguments.

8. Mathematics offers many opportunities for creative thinking, for exploring open ended situations, for making conjectures and testing them with data, for posing intriguing problems, and for solving nonroutine problems. Students in groups can often handle challenging

situations that are well beyond the capabilities of individuals at that developmental stage. Individuals attempting to explore those same situations often make little progress and experience severe and unnecessary frustration.

PARTNER ACTIVITY

- Trigonometry

  Turn to your partner and explain the sine, cosine, and tangent ratios. Use a diagram to aid in your explanation. Your partner will respond by using this information and writing three problems using these three ratios. You will then work the problems together.

- Geometry

  Turn to your partner and summarize what you know about two parallel lines cut by a transversal and the resulting angle relationships. Use a diagram in your discussion. Your partner will respond by writing three example problems. You will work the problems together.

THINK, PAIR, SHARE

Questions are an integral part of every teaching situation. We ask questions of our students to get them to draw conclusions, to make conjectures, to transfer previously learned material, to recall basic information, to advance their thinking to another level, and to check their understanding. We want our students to respond to questions ultimately to think, to understand, and to be able to communicate that understanding. Think, Pair, Share is designed to get students to react to questions and to interact with a partner. The concept is described by Canady and Rettig (*Teaching in the Block,* p. 75) as: "Each person in the classroom picks or is assigned a partner. The teacher asks students to enter Think Mode and then poses a question. After (at least) three seconds of 'wait time,' students are instructed to discuss possible responses with their partner. Students then 'Share' their responses as the teacher calls upon students randomly or solicits a volunteer."An alternate sharing technique is to ask students to write their responses on paper and/or draw diagrams. When called upon they read their responses and share the diagrams. This helps students become more concise in framing their answers. With two people writing responses, answers are clearer and more concise in using appropriate mathematical terminology. For more information concerning Think, Pair, Share refer to Canady and Rettig, *Teaching in the Block* (Eye on Education, 1996, pp. 75–78).

The following is adapted from the concept model shown in *Teaching in the Block* (p. 74):

THINK MODE

- The teacher instructs students to enter Think Mode.
- The teacher asks a question and establishes wait time.
- Students do not communicate to anyone (may write ideas down).
- No hands are raised.
- Students who finish early are asked to create a defense for their answers.
- Teacher signals Pair Mode.

PAIR MODE

- Students lean toward each other and communicate with their partner about the question.
- Students discuss a defense for their conclusions, rehearsing their response.
- Students use a "whisper voice" to discuss.
- Students refrain from shouting out.
- No hands are raised.
- Students come to agreement or they agree to disagree.

SHARE MODE

- The teacher indicates Share Mode with a signal.
- Teacher calls on students randomly or students raise hands to respond.
- Students refrain from shouting.
- Students share responses and defenses with the class in a variety of ways.

TEAM MODE (or SQUARE MODE)

- Teacher indicates Team Mode by a signal.
- Students lean toward the center of the group and do everything that is done in Pair Mode, attempting to arrive at a "team answer."

EXAMPLES

The following list of questions is suggested for use with Think, Pair, Share. Although listed by subject, some questions are usable in several categories.

- Algebra I

  How do you solve an equation in one variable?

  How do you solve an equation by completing the square?

  How do you find the intercepts?

  How do you determine the equation of a line given two points on the line?

  How do you determine if two lines are parallel?

  How do you determine if two lines are perpendicular?

  How do you find the slope of a line through two points?

  How do you determine if two lines are perpendicular if you are given their equations?

- Algebra II

  Given an equation representing one of the conic sections how do you recognize which conic the equation represents?

  How do you solve a system of linear equations (two equations with two variables)?

  How do you know if a quadratic equation has any real roots?

- Geometry

  Explain how to find the area of a triangle, a parallelogram, a trapezoid, a regular hexagon, or a regular polygon.

  How do you set up a proof?

  How do you prove that two lines are parallel?

  What does it mean to say that two polygons are congruent to each other?

  How can you show/prove that two figures are congruent?

  How can you show that two figures are similar?

  How can you tell if a triangle is a height triangle?

What is meant by the term "symmetry"? Give an example of things that have some type of symmetry.

- Trigonometry

  How do you prove a trigonometric identity?

  What would you do to find the height of a flagpole without climbing it?

  Why would the graph representing daily average temperature over a year model a sine wave? (Times of tides or times of sunsets or sunrises could also be used.)

- Calculus

  How do you solve an optimization problem?

  What information is important in sketching a curve?

  How do you find the area under a curve?

  What is the difference between average velocity and instantaneous velocity?

  What information can you determine about the graph of a curve given the second derivative?

  How do you determine asymptotes given the equation of the curve?

These are but a few examples of questions that could be used for Think, Pair, Share. We hope that these few will stimulate many many more for you to use with your students.

STUDENT TEAM LEARNING

There are three basic Student Team Learning techniques: Student Teams-Achievement Divisions (STAD), Teams-Games-Tournaments (TGT), and Jigsaw II. All three techniques have heterogeneous teams that work together to master a set of material and to receive team recognition for doing well academically. STAD and Jigsaw II use quizzes to do individual student assessment and to compute team scores. They use an improvement scoring system to allow students the opportunity to maximize their team scores and their contribution to the team. TGT uses academic game tournaments to assess individuals and to allow students equal chances to contribute to their team scores by having them compete against others of similar past performance. All of these methods work extremely well with middle school students and with students in high school enrolled in beginning math courses and/or in the ninth and tenth grades.

STUDENTS-TEAMS-ACHIEVEMENT DIVISIONS

"Students are assigned to four-member learning teams that are mixed in performance level, sex, and ethnicity. The teacher presents a lesson, and then students work within their teams to make sure all team members have mastered the lesson. Finally, all students take individual quizzes on the material at which time they may not help one another. Students' quiz scores are compared to their own past averages, and points are awarded based on the degree to which students meet or exceed their own earlier performances. These points are then summed to form team scores, and teams that meet certain criteria may earn rewards" (R.E. Slavin, *Student Team Learning in Mathematics,* in N. Davidson, *Cooperative Learning in Mathematics,* 1990, p. 71). This cycle of activity from teacher presentation to team practice to quiz usually takes two block periods. This method not only motivates students to help each other, but also encourages them to learn from one another and provides a method for organizing your classroom.

TEAMS, GAMES, TOURNAMENTS

Developed by David DeVries and Keith Edwards, TGT was the first of the Johns Hopkins cooperative methods. This method is a variation of STAD except that it replaces the quizzes with weekly tournaments in which the students compete with members of other teams to gain points for their team scores. The students are grouped in three person "tournament tables." They compete against others with similar mathematical backgrounds and level. There is a "bumping" procedure that changes the students' tournament tables each time frame (two-week time frames are recommended for block schedules) to keep the competition fair. The winner at each tournament table brings a preset number of points to the home team regardless of what table it is; this means that low-achievers (competing with other low-achievers) have the same opportunity to contribute points to their teams as do the high-achievers. Middle scorers (second place) earn a set number of points, as do the third place teams at each tournament table. All earn some points. The high-performing teams earn some type of reward or recognition. The dimension of excitement is contributed by the use of games. Teammates help one another in preparation for the games through study sessions, practice sessions, and discussions in their groups.

JIGSAW

Does this title make you think of a puzzle? This method is similar to a puzzle —the pieces come together to make a picture. Students are placed into groups of four or five and given a topic or concept to learn. In this first grouping, each person in the group becomes an expert on the same topic or concept. New groups of one expert from each group are formed and each person shares his or her expertise with others in the group. A variation is to assign each group a topic to become expert on and then as a group teach the other groups. All members in every group have an opportunity and responsibility to help others learn and to learn for themselves. In the Jigsaw II model, the teacher:

- provides an overview of the Jigsaw process to the whole class;
- assigns students to study teams;
- assigns each member of a study team to an expert group;
- gives a learning task (piece of the puzzle) to each expert group for mastery;
- reassembles study teams and each individual teaches the others in the study group his/her mastered task;
- assesses student learning. (Gunter et al., 1990)

EXAMPLES

- Geometry

  When beginning the unit on proving triangles congruent, assign groups. With a class of 30 make 6 groups of 5 students in each group and give each student in the group a letter of the alphabet (A, B, C, D, E). Group one is assigned the Angle-Side-Angle method; group two the Side-Side-Side method; group three the Side-Angle-Side method; group four the Angle-Angle-Side method; group five the Hypotenuse-Leg method for a right triangle; and group six the Hypotenuse-Angle method for a right triangle. Each group member becomes an expert on the method assigned to his or her group. New groups are now formed taking one member from each group. Place all of the students labeled A in a group, the B's in another group, and so on. In these new groups, each student teaches the others in the group his or her expert knowledge of proving triangles congruent. All the pieces come together to form a big picture of proving triangles congruent.

- Trigonometry

  Assign groups to investigate the effects of A, B, C, and D on the general equation $y = A \sin B(x - C) + D$ and $y = A \cos B(x - C) + D$. Group one investigates A, Group two B, group three C, and group four D for the sine function. Group five A, group six B, group seven C, and group eight D for the cosine function. Students are encouraged to use the graphing calculator in their investigation. Students then come together and listen to each group present its findings.

- For all classes

  Organize students into groups of three to check homework. Compare answers, discuss those problems that are not answered the same, make corrections, and state the reason the answer was changed. These can then be collected and recorded. The teacher can move about the room making contact with each group. This is a good opportunity to find those problems that the students as a whole had difficulty solving.

PROBLEM RELAYS OR PASS THE PROBLEM

Divide the class into five or six teams of five each. Be sure that the same number of students is on each team. Have the students sit by teams in rows (one student behind the other). The first person in each row is given a multi-step problem or exercise face down on the desk. When given the signal to begin, the first student turns the problem over, works only one step or procedure, and then passes the problem to the student sitting behind him or her. The second person also does only one step or procedure and passes it to the next person in the row. This continues to the end of the row. Students are reminded to work quickly but also to strive for accuracy. The last person in the row completes the problem and raises a hand or signals that his or her team is ready to submit its answer. The teacher checks the answer. The first row to finish with the correct answer wins points. The second team finished with the correct answer wins a lesser number of points. We give points to all teams who answer the problem correctly but in different amounts according to the order in which they finished with the correct answer. The largest number of points goes to the first team that answers correctly and the least number of points goes to the last team that answers correctly.

If the first person finished does not have the correct answer, have the last person start over and pass the problem up the row, having each student do one step or procedure that was done before. The teacher then checks the student's answer. If the team still has not gotten the correct answer, the team member can call for a huddle. They all get out of their seats and meet together to figure out the problem. They can check over their old work and/or do the problem again. This activity provides for individual work opportunities as well as team work and communication. Students find that it is important to show their work and to do it neatly so that the next person knows clearly what has been accomplished in the problem-solving process prior to receiving the problem.

These are some topic areas that lend themselves to this concept:

- Solving an inequality such as $2x + 3 < 9x - 5(x - 1)$ for x. This takes five steps. If there are five students for each team, each student will work one step of the solving process then pass it to the next team member to follow up by performing the next step in the solution process.
- For middle school, have students make a factor tree for prime factorization of composite numbers. The first student does a factorization and passes it on to the next student who does another factorization. This procedure continues until the number is written as a product of all primes.
- Solving any type of equation including an equation in one variable, two variables and isolating one of the variables, or a trigonometric equation.
- Proving a trigonometric identity. This can be a challenge!
- Finding a solution to a problem using the law of sines or cosines.
- Graphing an equation of the type $y = A \sin B(x - C) + D$. The first student would sketch the sine graph on a set of axes. All students would use the same set of axes. The next student would graph the sine function as it would be affected by one of the numbers representing one of the letters such as A. In other words, show the effect of A on the graph by changing the amplitude. The graph would then be passed to the next student and this graph would be changed to reflect the effect of another number in one of the lettered positions such as B. This would continue until the complete graph had been graphed with all of the transformations shown on the same set of axes. Students must know the significance of each letter on the transformation so that when the graph is passed to them they will recognize what has been changed and what is left to be changed.

  These transformations can be applied to many graphing situations. This technique works well with graph review during the first few days of Calculus to summarize graphing strategies in all types of equations.
- Give the students the coordinate of a point on the terminal side of an angle and ask them to find all six trigonometric ratios for the angle.
- Give the students the quadrant in which the terminal side of an angle lies and one of the trigonometric ratios for the angle and have them find the other trigonometric ratios.

- Give the students the lengths of two sides of a triangle and the measure of the included angle and then ask the students to find the measure of the other side and of the other two angles.
- Set up a proof complete with the diagram and the given and prove statements. Also include the two-column proof set-up with a column for statements and a column for reasons. The proof is then passed through the sequence of students on a team and each is to fill in the statements and reasons by doing only one statement and its corresponding reason. Then they pass it on to the next team player until the proof is complete. An alternate form of this strategy is to fill in some of the statements and reasons and leave some blank for the students to fill in. Another suggestion is to give them all of the statements without any reasons and have them put them in the correct order.

USING MEDIA CENTER RESOURCES

The media center specialists in your school can be a source for creative projects and ideas for the mathematics classroom. They can provide your mathematics department with a list of all math-related books in the library, as well as software, videos, compact discs, slides, and information sources on the Internet. They are often delighted to get suggestions for books and materials that are math related. We have provided a list of Web sites at the end of this book. We have used or visited each of these sites; they will assist you in your search for ways to use the Internet for instructional purposes or for idea searches. Some of the activities we have used follow.

- Scavenger Hunt

  Years ago, the scavenger hunt was a popular game played at many preadolescent and adolescent parties. All of the party attendees were put into groups and given a list of objects to collect from friends and neighbors in a given amount of time. The groups that had collected the most objects from the list were the winners and were awarded a prize. In the classroom, the students are placed into groups and given a list of math questions or problems to answer. The students use the media center (library) and/or the Internet to answer the questions. They can be given a class period in the media center to complete the activity or they can be given a deadline for later completion of the hunt. The group that has the greatest number of correct answers at the deadline is the winning group. Prizes such as crazy pencils, stickers, candy, bonus points, or special ribbons or buttons can be awarded. *Discovering Geometry* has several scavenger hunt activities on pages 122 and 216. There is also "The Great Mathematical Scavenger Hunt" in *The Mathematics Teacher*, NCTM, October, 1980, p. 513.

- Individual projects such as: *What is Chaos?; Why Does a Golf Ball Have Dimples?; Math in Architecture.; Who Was Euclid?; Who Was Pythagoras?;* and *Math in Nature: Soap Bubbles and Math.* (See the section on projects and culminating activities.)
- Group project or a class project. Split the class into four or five groups and assign each group a topic to research and to report back to the class as a whole. Select a time period such as the Renaissance and assign one group to research art and music during the Renaissance. The second group would research the historical events during this time period; the third group would research the scientific discoveries and inventions and their impact on society during the Renaissance; and the fourth group would research the literature of the same time period. Another example: assign each group a civilization such as Babylonian, Mayan, Arabian, Egyptian, Chinese, or Greek. Ask each group to find the contributions made by the assigned civilization to mathematics and to present the major contributions using a visual such as a poster or model.
- Assign readings in journals and ask the students to write a summary.
- Research what new mathematics were discovered in the last 10 years.
- Career Project. Students are assigned a career or profession that has strong connections to mathematics. They research it and write a job description: the necessary background and education required, the average starting salary, and where to go for more information. The students then give a short presentation to the class. All of the reports are collected and made into a math career library to be kept in the classroom for students to use.
- Have students consult the *Almanac* to obtain information to form a problem. Our prealgebra students chose the Olympics as their theme the first grading period. Each student chose an event that used time to decide the winner. They watched the events on television and recorded all the times of the winners in their category. We then went to the media center and used the *Almanac* to gather information about each student's event and to record all the times of the winners since

the records began to be kept. Each student then used his or her information and predicted the winning time in the event for the year 2000. The students put their information on a poster along with a colorful picture of the event in progress or the winner's photo. In addition, they made a table of the years and times of the winners in their events and placed them on the poster along with their prediction and a justification for the prediction.

Other media resources include *Guinness Book of World Records*, recorded music, speeches, poems, plays, novels, and drills. Filmstrips, pictures, and transparencies help keep students' attention and make the lesson "come alive." Seize the opportunity to use the media center and its available resources whenever the curriculum renders it appropriate.

GAMES, PUZZLES, AND RECREATIONAL MATHEMATICS

Many teachers are reluctant to use games in the classroom because they too often perceive them just as games and not as valuable teaching and learning tools, or as "hooks" or motivators. Games and puzzles provide variety and fun in the mathematics class and can be used for test review, skill and drill, or as a tactic to involve students in learning math.

A personal experience of one of the authors from this last year casts games in an interesting light. The author taught two different levels of Geometry classes. One was an accelerated class of all ninth graders, many of whom were classified as gifted. The other class was composed of tenth and eleventh graders about half of whom were classified as college-intending. Both classes had finished units on logic and had worked some logic puzzles. We had talked about using strategies and making charts to help solve this type of problem. Some students dove right in and seemed to have had experience in working puzzles or were very intuitive. Others did not have any notion, before our discussion, of how to start solving the puzzles. This piqued my curiosity about their past experiences playing strategy games or cards. The students in each class were asked, "How many of you grew up playing cards or board games at home?" In the accelerated class, all of the students raised their hands and in the other class only 4 of 24 students raised their hands. What an interesting contrast—was it coincidence or is there something to it? Of course, this discussion and conjecturing is for another time and place but it did make this author think: Would playing strategy games and card games improve their deductive skills?

Using games and puzzles developed to enhance the learning of mathematics provides variety and fun for the students. They are good for problem solving, analyzing, synthesizing, retention and review, logical reasoning skills, and listening and communication skills. The following games were selected because they have been used in the mathematics classroom successfully and because they require little or no money and little preparation time for the teacher.

WHEEL OF FORTUNE

This variation of the popular television game "Wheel of Fortune" requires very little time to prepare, and usually takes 15 to 20 minutes of class time. It can be used as part of a review for a test or even a warm-up activity on days when students need practice on concepts and no new material will be introduced.

To begin play, think of a definition or a question related to the subject matter. For example, "What is the relationship between the slopes of two parallel lines?" Outline the words of the question on the chalkboard or on the overhead using dashes to represent letter blanks. Each student in the class is an individual contestant (or you can put the students into groups of two or three).

Start the game by writing a problem on the board. The first person to raise his or her hand and give the correct answer gets to guess a consonant. If this letter appears in the message, it is written in the appropriate blank(s) and the student's name is written on the board. Write another problem on the board and follow the same procedure. For a student to buy a vowel, he or she must answer a second question correctly. To win the game, a student must correctly guess what the question says and also give the correct answer. The students whose names are on the board are the only ones who qualify to solve the puzzle, and they may try to solve it at any time during the game. The teacher may instruct puzzle solvers to raise both hands to signify their intent to solve, or each student may be given two sheets of paper of different colors, raising one color to answer the question and other color to solve the puzzle.

Teachers will find that they can make up most or all of the mathematics problems and answers in their heads, depending on the complexity of the subject matter. However, it is a good idea to have a bank of problems and answers done ahead of time to maintain momentum. (Idea adapted from *Mathematics Teacher*, December, 1987)

MATH JEOPARDY

This game is a variation of the popular television show "Jeopardy." All you need to play this game is a basic understanding of the television show, six sheets of each of five colors of construction paper, a marker, and a chalkboard. This game is ideal for any math class in middle school or high school. It takes about 20 minutes of teacher preparation time and at least 30 minutes to play. It can even be continued to another day if you need to stop play in the middle of a game.

To prepare, the teacher selects five categories and writes five questions for each category. The questions should be of varying difficulty in each category.

Write each category name on a different color of construction paper and tape it to the board in one row at the top of the board. Use the remaining paper to write questions on one side of the paper and a dollar amount on the other side. Be sure to use the same color of paper for questions in the same category. Tape these, with the dollar amounts showing, in columns under the category headings making a grid of solid colored columns. Make sure the dollar amounts are in increasing order. The game board is now ready for play.

To begin play, separate the class into two teams of students of equal ability. The groups then choose a captain and they also select a name. Write the name of each team on the board. Use these to keep score. After deciding which team goes first, the captain of that team selects a team member to start play for that team. The team member selects the category and dollar amount for the question. The teacher reads the statement after turning over the dollar card from the board. The student, without any help from teammates, has a specified amount of time to answer. If he or she answers it correctly, then the captain selects a second team member to go next. This team member also selects the dollar amount and the category for his or her question and play proceeds as before. The maximum number of consecutive correct responses for one team is three and then play goes to the other team for the same number of consecutive correct responses. However, if a person misses a question, then the question goes to the other team, whose members may confer and answer. If the answer is correct, that team begins its turn and the other team loses its turn. If the conferring team misses the question, play goes back to the original team. That team starts fresh with another turn and play continues.

To keep track of the scoring, place the paper with the dollar amount on it under the name of the team on the board and keep a running total of dollars from both teams.

After all questions have been answered and scores totaled, "final jeopardy" is started by announcing the category. Each team then confers for one minute to decide the amount of its wager. They must write the wager down on paper and put it on the board by the team total. The question is then read and the teams are given one minute to write their answers and give it to the captain. The teacher reads each answer and totals the scores. (Add a drum roll or two for dramatic effect!) The game is over and the winners are announced and congratulated.

There are many other games and puzzles to use as models for math activities including Bingo, Trivial Pursuit, Pictionary, Balderdash, Outburst, and Scattergories. There are also some excellent games and puzzles in the *Mathematics Teacher.* Some of my favorites are: *Petals Around the Rose,* December, 1978, p. 753; *I Have a Number,* October, 1988, p. 650; *I Have...Who Has...?"* October, 1980, p. 504; *Algebra Tic-Tac-Toe,* January, 1983, p. 35; and *Geopardy,* December, 1987, p. 722.

ENRICHMENT, CREATIVE PROJECTS, AND CULMINATING ACTIVITIES

"Enrichment activities play an important role in the mathematics program. When enrichment is an integral part of their work in mathematics, students of all ages and abilities learn to work in mathematics and to appreciate mathematics as a living, useful, interesting subject"(*Mathematics Teacher,* September, 1990, p.451). As teachers of mathematics we want our students to be able to do mathematics and at the same time to see its beauty as a living, growing, useful, and interesting subject. Not only should we strive to do this with each lesson, but we should also individualize enrichment for each student allowing for individual talents and special interests through group and individual projects and enrichment activities. These can also be used as a form of assessment and/or as a culminating activity at the end of a unit, or as a "hook" at the beginning of a unit.

Using projects is one way to stimulate students to be creative, to use what they already know and to seek to know more. "A mathematics project consists of all the effort expended in solving a problem, exploring an idea, or applying a mathematical principle—that is, the initial planning, the study, the exhibit, and the write-up or report. It should develop a new mathematical concept or theory, show the relationship of a mathematical idea or principle to some other branch of mathematics or science, or demonstrate the application of a mathematical idea or principle. The exhibit should be an attempt by means of drawings, graphs, models, pictures, words, and so on, to tell the viewer briefly the student's idea of the mathematical concept or principle, to show its use in explaining some biological or physical phenomenon, or to answer a question posed by the author of the project. The exhibit may be a collection, an experiment, or a working model. It may be a repetition of an earlier experiment or an entirely original work."(*Mathematic Projects Handbook,* NCTM, 1977, p2.)

Many aspects of mathematics are concerned with applications such as art, music, news, cartoons, architecture, nature, and photography. Other applications deal with how mathematics is used in an occupation or how it has affected physical theories, political thoughts, the social sciences, or ancient cultures. All of these are starting places for ideas for projects. (Other resources for topics can be found in NCTM's *Mathematic Projects Handbook,* pp 8-18.)

The very best projects grow out of students' interests or natural curiosity; therefore, the students should choose their own topics with their own special interests and talents in mind. To stimulate topic ideas, show the Walt Disney video *Donald in Mathmagicland*. Use the pause feature to stop the video and discuss particularly interesting information you think the students will enjoy. A variation is to use a section of the video that has a preselected focus. For example, if the focus is mathematics in art, show only the film clips relating math to art.

If you do choose to see the entire film, follow it up with either a class discussion, small group discussions, brainstorming activities, or web activities on the chalkboard. The students will generate enthusiasm and become curious about answering self-initiated questions.

Our students are assigned a project or activity each grading period. Some assignments are for enrichment while others are more for investigations. Some students choose to build models, to make drawings, or to use their artistic talents. Others who enjoy using the computer find interesting ways to utilize it in their projects. Try to find avenues for all students' talents and interests when selecting the projects and activities, especially because they are required. The activity that we select each six weeks is related to what we are studying during that time frame or is an extension of mathematics to the "real world." One assigned project is interdisciplinary.

As we anticipate the beginning of a new school year and new classes, we reflect on the strategies used in the past and note those that worked well and those that did not work so well. We then select projects and enrichment activities for the coming year. This past year the following projects were assigned to our geometry classes. A brief description for each of the assignments gives an overview of the activity. The students are given specific handouts for each of the activities including guidelines and instructions.

FIRST SIX WEEKS: INTERVIEW

Students are to interview someone to find how math is used on the job and how much math education is required. From information gained during the interview, students make a poster. The poster has four parts: (1) the title of the profession; (2) a visual representing the occupation; (3) a paragraph or two relating the information about the occupation; and (4) a typical mathematical problem from the occupation. Because of the varying complexity of the mathematics involved in solving these occupation-related problems, students are not expected to understand their solutions. For example, a geometry student would not understand the calculus involved in a problem related to aeronautical engineering. On the back of the poster the student must put the name, address, and title of the person interviewed as well as his or her own name, class period, and date ("Career Posters," Peggy Tibbs and Janette Jordan, *Mathematics Teacher*, Sept. 1994, pp. 410–411).

Many of the students interview their parents and even include pictures of them on the poster. This gives the parents the opportunity to communicate with their children about the kind of work they do and about how they use math in their professions. Student interviews have included talking with a delivery person from UPS, a carpenter, an engineer, a psychologist, a landscaper, an optometrist, a business manager, an accountant, a variety of doctors, a self-employed music store owner, a cabinet maker, a chemistry professor, a city planner, and a Spanish teacher. This information provides a list of potential guest speakers and sources of authority on different subject matter.

SECOND SIX WEEKS: CREATIVE PROJECT "PICTURE WORTH A THOUSAND WORDS"

Students chose a feeling, emotion, or descriptor of a personal nature, which they will use to create a design. The picture is composed of an arrangement of construction paper cutouts of geometric shapes placed on a sheet of construction paper. The design is to visually represent the descriptor they have chosen. On the reverse side of the construction paper design the student attaches a written description that states the theme of the work, and then explains how the picture expresses or represents the theme. The student should include the relationship between the shapes and the colors chosen and discuss how these reflect the theme. The student should write in complete sentences and full paragraph form.

THIRD SIX WEEKS: *FLATLAND*

Students read the book *Flatland* by Abbott. On the due date the students are given a quiz to verify that they have read the book. This is followed by group discussion. Groups of four or five students are then asked to pull a topic related to the book from a hat and to perform a skit addressing the selected topic. Examples of topics are being a female in Flatland or the hierarchy of the inhabitants of Flatland. Colored paper, markers, scissors, tape, and so on are available for the students to use to make props and/or simple costumes. Students interested in the social sciences and drama enjoy this activity because of the opportunity to be creative. This activity is well suited for the last two days before the Christmas or winter holidays.

FOURTH SIX WEEKS: SCAVENGER HUNT

The scavenger hunt can be one designed by the teacher or it can be one that is predesigned such as the scavenger hunt in the October 1980 *Mathematics Teacher* (p. 513). This particular scavenger hunt is an ideal project for using the Internet for research because it is a real challenge! The hunt can be assigned to individual students, to partners, or to small groups. We give points to all students who participate in the scavenger hunt and complete it. The largest number of points goes to the first group to finish and to have the correct answers, the next group finished gets the second highest number of points, and so on.

FIFTH SIX WEEKS: SPECIAL SELECTIONS

The students select two topics or activities from a list of project topics provided by the teacher. The students are asked to use a visual and to include a one-page report with a bibliography for each selection. The list might include: Math in Architecture, Math in Nature, the Golden Ratio, Math in Music, or Math in Quilt Design. This project assignment may be preceded by showing Walt Disney's *Donald in Mathmagicland,* which is full of ideas for projects.

SIXTH SIX WEEKS: PLATONIC SOLID MOBILE

Students are given a brief history of the Platonic solids and then are provided the patterns for the Platonic solids. They are instructed to make mobiles using the solids that they construct from the patterns. Accuracy and creativity are emphasized. The next step is for them to find the necessary dimensions to calculate the surface areas and volumes of the constructed Platonic solids. If a student cannot find a formula for either the surface area or the volume of a solid, the student is to describe a method that may be used to approximate the measurement in question. After the students turn in their projects and they have been graded, the mobiles are hung in the room from the ceiling.

USING A THEME

While teaching Algebra I, Part I, we chose a theme for the projects and enrichment activities. That year happened to be an Olympic year, so the Olympics became the theme. For their first project the students selected an Olympic event that involved scores using time or distance, such as the women's 100-meter sprint or the men's pole vault event. Television guides were provided in the classroom and the students were to record the days and times that the selected events would be televised. This project required the students either to watch their events and record the winners and their times or distances or to obtain the data from a news source.

After this data collection the students were taken to the media center (library) where the media center specialist helped them find sources that contained winning times by year for the selected events. These past scores were recorded for use in the next phase of the project.

The students took this data and made a line graph, or scatter plot, placing the years on the horizontal axis and the time or distance on the vertical axis. (The students could input this information on to the graphing calculator and copy the calculator's graph. A linear or other type of regression could be done and included in the project.) Using the graph information the students were required to predict the next time or distance for the event and then to defend their prediction. Now students were ready to put it all together.

Using a piece of poster board, each student made a poster that included: a picture showing the selected event; the graph of all of the years and scores for each; the student prediction and reasons explaining the prediction; the name of this year's winner; and the winning time (or distance). The poster was titled with the name of the event. No pencil was used and all writing on the poster board was word-processed. When we assigned the project we made a mockup of a poster to show the students what was expected from them. They responded by turning in some of the very best projects we had ever received.

Many times the reason inferior work is turned in is due to the students lack of experience or to what they perceive as acceptable. When they can see a project from a past year's class or one that the teacher has done, it becomes very clear what is expected.

Other themes that could be used are space exploration (Mars on the World Wide Web!!), new math discoveries (chaos or fractals), math in the newspaper, being a consumer of goods, or the stock market. Selecting a theme and making it "real" is a way to help students to view mathematics as an important part of the real world.

5

USING TECHNOLOGY IN THE BLOCK SCHEDULE

WHY USE TECHNOLOGY IN THE MATHEMATICS CLASSROOM?

As technology becomes part of our everyday lives, it becomes a part of the education of our children. Television provides a way for us to see world events as they happen. It enables us while we are in our own living rooms to see others in their own cultures, worlds away in distance. The age of technology and information worldwide is here. The "super highway," laptop computers, networks, and fax machines are a part of our lives and have been our means of becoming part of a global economy. Companies have branches located throughout the world, which means we are doing business with people from other cultures and that we are in competition with them. Technology has allowed us to "reach out and touch someone" as often as we like. Technology and being part of a global economy are making demands on the education we are providing for our children. What impact does it have on education and how can we prepare our students for the future? What can we as teachers do to insure that our students are prepared for a future using technology?

"Computers, in short, are tools for restructuring virtually every aspect of American education" (Edward Fiske, *Smart Schools, Smart Kids, 1991, p.161*). How can we in education use this tool and how does its use prepare our students for a world of technology and information? Computers are powerful tools for working the curriculum of schools away from rote learning and toward problem-solving skills that students will need in the workplace of the future. Computers are inherently social devices—a document or a computer screen is public property. They enhance cooperative learning. Students work together naturally on computers. Current trends are toward the use of computers not as a means of delivering instruction, but as a means of controlling and managing the interaction of all sorts of other technologies: written documents; film; videos;

audiotapes; CD-ROMs; information from outside databases; and the like. The sky is the limit, both literally and figuratively. "Technology is a license to think creatively," said Eugene Hetrtzke, Superintendent of Central Kitsap School District in Silverdale, Washington (Fiske, *Smart Schools, Smart Kids,* 1991, p. 160).

Incorporating technology into the classroom encourages and provides a means for teachers to interact and communicate with one another. They also become "managers of instruction, not presenters of information"(p. 152) and function as a team member sharing ideas and communicating those ideas. Technology encourages students to learn on their own and in groups, and to play an active role in their own learning. "Technology should be employed to manage learning as well as diagnose, present and evaluate it."(p. 151)

In our everyday lives, we are touched by technology; for example, when we use a charge card, or go the bank and use a cash flow card to withdraw money, or when we call the bank to get a list of deposits and/or withdrawals, we use technology. We use cards when we put gas in our cars and then pay the amount registered. When we go the the library to find a book and use the card catalog we use the computer to find the location and whether the book is on the shelf or checked out. When we go to the paint store to select paint, there is a computer that takes an image of a room and paints it the color we selected. We can get a picture right there in the store. When we redesign a kitchen or a house, we can go to the design center and use the computer to put everything together and get a picture of what it will look like. As consumers we need a basic understanding of what we can use the technology for and how it benefits us.

This need translates into using computers as a tool for teaching and learning. There is a fascination with interesting ways of teaching, and the computer is used as an "amplifier" which makes teaching more powerful. It may be a catalyst for better education. From the students' perspectives, computer-based learning might challenge students to think about what they know and to challenge their misconceptions. (Fiske, *Smart Schools, Smart Kids,* 1991, p. 150)

When the composition of our society changes and cultures blend to form a new society, and as the family changes, the needs of our children change. Education must become part of the solution to the problems that arise. As we shift from an industrial society to an information society, we need to use technology as part of the restructuring of the education process. There is overwhelming evidence that education is in a state of change. These changes manifest themselves in the basic institution of education by way of free choice, in the role of the individual school through site-based management, and in the very essence of day-to-day education in the interaction between teacher and student. We believe that for a significant change to occur, all three must happen and must have the support of educators, students, parents, and the communities schools serve. Our schools must meet the needs of society and prepare our students for a future in that society.

THE STANDARDS AND TECHNOLOGY

To incorporate technology into the classroom you must first have access to the technology. Calculators, computers, courseware, and manipulatives are necessary for good mathematics instruction (NCTM, *Standards*, 1991). Teachers can no longer rely solely on the chalkboard, paper-and-pencil and textbook. It is also obvious that simply providing the technology is not sufficient; teachers need to know how to use the technology and how to integrate it into quality mathematics instruction. Teachers must have support from their administrators, parents, and school boards in providing appropriate training, workshops, and access to resources. These cost money. Money is a barrier to change and to reform because most schools are "surviving, not thriving" in the face of cutting or maintaining local budgets. For success to occur, reform requires a commitment of time and resources. We must find the time and the resources for our students.

"The major influence of technology on mathematics education is its potential to shift the focus of instruction from an emphasis on manipulative skills to an emphasis on developing concepts, relationships, structures, and problem-solving skills" (*Mathematics Teacher,* April 1985, p. 244). It is assumed that all high school students will have access to graphing calculators and that they will have the opportunity to use computers in mathematics. Also, there will be a computer and an appropriate projection device available for instruction in every mathematics classroom.

In the *Standards* and books and articles discussing the *Standards* there are important statements that give us the "big picture" of what the role of technology should be. They are listed here under the appropriate headings.

- Middle School Curriculum
  - "Calculators should routinely be available to students in all activities associated with mathematics learning, including testing. Students should be taught to distinguish situations in which calculators are appropriate aids to computation from situations in which mental operations or paper-and-pencil computations are more appropriate.
  - "Mathematics in middle grades should emphasize the development of 'number sense'—the intuitive feeling for the relative sizes of numbers that is essential in skillful estimation, approximation, mental arithmetic, and interpretation of results for reasonableness.
  - "Calculators and computers can be used to teach iterative procedures for solving significant problems before traditional formal methods are presented.

- "An introduction to statistics should include extensive gathering, organization, and presentation of data. The manipulation and examination is aided by computer analysis and graphing software.
- "Middle school mathematics should take advantage of the visual display capabilities of computer graphics that support and underscore the importance of informal geometry objectives. Transformations, mensuration formulas, and spatial visualization can be vividly illustrated by using computer graphics.
- "Increased emphasis should be placed on such nontraditional methods of problem solving as organized lists, guess and check, geometrical sketches, and successive approximations, all of which are made feasible by calculators and computers.
- "Computer programming experiences that introduce the concepts of variable and function should be provided. These experiences should help prepare students for the study of algebra."

NCTM, *Standards,* 1991.

- High School Curriculum
  - "Computing is changing calculus and its traditional prerequisite subjects. Computer symbolic systems, graphics, and numerical analysis software make student mastery of manipulative procedures less important and, at the same time, offer dynamic tools for the illustration of fundamental concepts and processes. Such changes should be considered as the influence of calculus on high school mathematics is assessed.
  - "The geometry of three-dimensional space, trigonometry, vectors, coordinates, and transformations are all made more accessible with computer assistance through visual displays and complex calculations.
  - "The skill objectives of algebra must be reassessed to identify those procedures more easily done by computer or calculator. The properties of elementary functions are still important for modeling quantitative relationships, but proficiency in many familiar computational processes is of little value.
  - "In many topics of high school mathematics, computers and calculators can be used to discover and test principles and methods. For example, concepts and theorems can be illustrated numerically

> and graphically to develop sound understanding before formal proof is attempted."
>
> NCTM, *Standards*, 1991.

Teachers of mathematics should encourage and accept the use of computers, calculators, and other technology. Technology is an effective tool used in discovery lessons as well as in cooperative group activities. Students should be able to use calculators or appropriate software to solve problems that require tedious computation or repetition. Also, assessment models should be designed to use technology.

To quote Mary Ann Matras (*Mathematics Teacher*, February 1991, p. 86–87):

> When do we begin to use technology in the mathematics classroom? Whenever and wherever the technology allows students to do mathematics more efficiently and effectively. When do we end our explorations of the use of technology in the mathematics classroom? Not when we have learned to do all the old things a little faster but when we have found new ways to do mathematics and ways of doing new mathematics while using the technology to its fullest potential. And that quest should keep us and our students exploring and learning well into the twenty-first century.

GRAPHING CALCULATORS

The graphing calculator is making a tremendous impact on math instruction and on the learning of mathematics. It has a multitude of uses and is relatively inexpensive when compared with the cost of a lab full of computers and the software needed to work with a class of students. In addition there are numerous graphing calculator activity books available for almost every math subject taught from middle school through Advanced Placement Calculus. The *Mathematics Teacher* and the *Mathematics in the Middle School* journals frequently provide activities and resources for graphing calculator lessons.

Five years ago at our high school we purchased our first classroom set of graphing calculators. We got the first set to use in the Advanced Placement (AP) Calculus course. We were aware that it would become a natural part of the course of study and then a part of the AP Exam. This need justified the purchase. In a department of 18 math teachers, each teacher has a classroom set. We also have six overhead projector graphing calculators to share. We collaborate and explore as teachers sharing ideas, successes, and failures with the calculator. We have found value in using the graphing calculator to help students discover, investigate, and analyze data; make predictions; observe patterns; formulate equations; and actually "see" mathematics in action.

Many students have purchased their own graphing calculators because they know the calculator is an integral part of mathematics instruction. Some of these same students have become experts on the calculator and are eager to share their insights with the teacher and with their fellow classmates. The graphing calculator is a stimulus for students to question and to problem solve. It also stimulates interaction and discourse among students while doing mathematics.

We updated our curriculum this last year and included the use of the graphing calculator in the curriculum. This was a major step to elevate its importance as an instructional tool and as a learning tool.

We have used the graphing calculators in the traditional class period and then in a block schedule. The block schedule provides the time to use the calculator to discover, explore, and investigate, and then have time for students to discuss those findings, respond, and then to formalize what they have learned. The block schedule allows for the student to get the "big picture" and then to fine-tune it for the details.

The three areas of instruction that can be impacted the most through the use of graphing calculators are function analysis, data analysis, and solving equations. These three topics flow through all algebra courses, geometry, precalculus, calculus, and probability and statistics. With the use of the calculator students are better able to understand concepts and have an appreciation for functions in real applications.

According to NCTM's *Algebra in a Technological World, Curriculum and Evaluation Standards* (1996, p. 2), using the graphing calculator or other graphing tools influences mathematics content in a technological world in these ways:

- They allow a ready visualization of relationships.
- They allow the accurate solution of equations and inequalities not possible through symbolic manipulation alone.
- They provide numerical and graphical solutions that support solutions found using algebraic manipulation.
- They promote exploration by students and their understanding of the effect of change in one representation on another representation.
- They encourage the exploration of relationships and mathematical concepts.
- They promote "what if" modeling of realistic situations.

EXAMPLES

The following are examples of activities that highlight features of the graphing calculator.

MIDDLE SCHOOL MATH OR PREALGEBRA

- As students learn the order of operations and they begin calculations of either long sets of numbers or those that involve many operations, the graphing calculator is preferable to a regular calculator. The graphing calculator screen shows exactly what the student has entered, thereby making it easier to identify trouble spots on getting an incorrect answer. This is both beneficial to the teacher and to the student. The students are able to compare what they have entered into the calculator to what the teacher has entered on the overhead graphing calculator. For example, when asked to square a negative number such as -3, students who enter -3^2 will get an answer of -9. The calculator helps students see they should be squaring the whole quantity and not squaring 3 and taking its opposite. This emphasizes the role of parentheses.
- After students have collected data or are provided with data collected from survey results, the data can be used to make predictions. Height has been compared to shoe size; statistics from football, baseball or basketball teams; winning times for Olympic events over a number of years and the results of the United States Census since it started. Keeping in mind that we only want to use two variables, these results are given to the students in a table. The students are asked to plot the data on the graphing calculator and then answer guided questions. As the number of units on the horizontal axis increases, what happens to the number of units on the vertical axis? Do they both increase together? Do they both decrease together? Does one increase and the other decrease? Ask students to make predictions regarding the quantities in the table.

ALGEBRA I

SLOPE

- To investigate the concept of slope and how lines differ according to their slope, give the students two grids with four to five linear equations in slope intercept form for each grid. Make the slopes in one group have positive values in increasing order and the slopes in the other group have negative values in increasing order, keeping the y-intercepts the same for all equations. Students graph each of them on the graphing calculator and copy the graphs with positive slopes on one grid and the graphs with negative slopes on the other grid.

Students then are asked to make conjectures using these graphs and/or use the graphs to respond to questions.

- If the value of the coefficient of x is a positive number, describe the line on the grid.
- If the value of the coefficient of x is negative, describe the line on the grid.
- As the coefficient of x gets larger (or smaller), what happens to the graph of the line?
- Write the equation of a line that is between two lines in each group. Designate which two lines you want the students to use. They are then to check their solution on their calculator by graphing it.
- Write the equation of a horizontal line.
- Write a short description of all your findings.

Extension: This also can be adapted for investigating the effect of b on the graph of the linear equation $y = mx + b$.

◆ This is a culminating activity to be used after students have learned the slope and y-intercept of a line in $y = mx + b$ form. Using the overhead graphing calculator enter the equations of lines. (The number of equations you enter will depend on which calculator you are using. Some calculators will allow only four equations to be entered at a time and then graphed.) The students are not to see the equations. Show the graphs one at a time on the overhead screen and ask the students to write the equation of the line on the screen. After completion, lead the students in a discussion about the clues used in arriving at the answers. This is a good assessment activity.

ABSOLUTE VALUE

◆ To investigate the absolute value function provide the students with pairs of equations in which one equation is linear and the other is the absolute value of the linear equation. Examples are:

$y = 2x$	$y = \lvert 2x \rvert$
$y = -3x$	$y = \lvert -3x \rvert$
$y = x + 1$	$y = \lvert x + 1 \rvert$

Students are asked to graph one pair on the graphing calculator and then copy the graphs onto a grid provided on the worksheet. They continue this for all pairs of equations. When they have completed the graphs, ask them to describe how the graph of the absolute value function compares to the graph of the original function. Describe all of the values of y for the absolute value functions.

This activity can be extended by placing the absolute value about different parts of the linear equation such as $y = |-3x| + 2$. Students can investigate the effects of each change using absolute value.

COORDINATE GEOMETRY

- Using the graphing calculator and the DRAW menu, create a square with each side 20 units long and 1 corner at (4,1). Find the coordinates of each vertex. Can you find a second square using the same description? If so, state the coordinates of the vertices.
- Create your initials and list enough coordinates so that your partner can reproduce them exactly as they are on your screen.
- Draw a right triangle with one vertex at the point (5,6). Write down the coordinates of the other two vertices. How do you know it is a right triangle?
- Draw a rhombus. List the coordinates of the vertices. What are the coordinates of the point of intersection of the two diagonals?

ALGEBRA II

- To evaluate a function at a given value of x begin by writing an algebraic expression in terms of x on the overhead with a table of values chart under it. Use two columns with one labeled x and the other labeled y. This activity has three parts. Part one requires the student to use the calculator to find the value of a given expression, for example at $x = -2$ by substituting -2 into the expression for x and calculating its value. Part two requires the student to graph the equation on the graphing calculator and then use the TRACE function to find the values of y, for example, at $x = -2$. Part three requires the student to use the ask feature of the TABLE function on the graphing calculator to find corresponding values of y, for example, at $x = -2$. Repeat this process for other values of x.
- Use the graphing calculator to solve a quadratic equation. After students have solved quadratic equations by factoring, the following activity gives them a visual representation of the solutions. Provide five equations, four of which can be factored over the set of integers and the fifth equation which is not factorable over the set of reals, for

example, $x^2 + 1$. A grid for each equation should also be provided. Students are to graph the equations one at a time and copy the graph onto the grid. They are also to solve each by factoring and to write down the solution(s). Then they are to state whether the graph crosses the x-axis and if it does, where. Ask students to make a conjecture regarding the solution to the quadratic equation and the roots of the graph of the equation.

Extension: Provide several quadratic equations and ask the students to find the roots, if they exist. Also ask them to compute the discriminant ($b^2 - 4ac$) for each. Is there a connection between the value of the discriminant and whether or not the equation has real roots?

TRIGONOMETRY

- Verification of identities can be done using the graphing calculator by taking the trigonometric expression on each side of the equals and entering each into separate $y =$ slots on the graphing calculator. Students then are asked to be sure to adjust the calculator to graph in the sequential mode. If there appears to be only one graph on the screen, this indicates the two graphs are identical. Ask students if this is enough to confirm that the two are equal. How can the exceptions for x be found if there are any?
- After students have learned the effects of a, b, c, and d on an equation of the type $y = a \sin b(x - c) + d$, use the graphing calculator for an assessment activity. The teacher enters several equations of this type into the calculator and, by using the overhead projector connected to the graphing calculator, shows each graph one at a time and asks students to write an equation for each graph. To check if the students are correct, their responses are entered into the overhead calculator and graphed with the original to see if they are the same. This works especially well if you are lucky enough to have two overhead graphing calculators. One can be used for the original graphs and the second one used for the student responses, which can then be compared side by side on the screen or wall.
- To investigate the domain and range of a function that results from the sum of two functions, the students and teacher enter each function into a $y =$ slot and then use the Y VARS button to enter the sum of y_1 and y_2 into the y_3 position. First graph all three graphs sequentially noting high points, low points, and where the new graph y_3 crosses the x-axis. Discuss what appears to be the domain and range

of the y_3 function. Next go to the TABLE on the calculator and compare the y_1 and y_2 values with y_3 values. Ask students to write a statement about this relationship and, at completion, discuss the statements and formalize a conclusion. This also can be done for a function formed by either a difference, product, or quotient of two functions. This visual approach to domain and range offers a concrete example to students.

This activity also can be adapted to functions other than trigonometric ones.

PRECALCULUS

- To investigate exponential functions of the form $y = a^x$, use the graphing calculator. Provide a worksheet with three grids, two to be used with sets of four equations and the third for experimentation. In one set of equations, the base *a* should be whole numbers greater than 1 in increasing order. In the second, set *a* should be values between 0 and 1, including 1, in increasing order. The students graph the equations from the first set on the calculator and then sketch them on the first grid. They then graph the equations in the second set on the calculator and sketch them on the second grid. Then they write responses to these questions: What happens to the curves as the bases increase in set 1? In set 2? When a = 1, what is different about the graph? Does it belong in our group? Why or why not?

  The third grid can be used for further graphs with different bases if the students need more graphs before arriving at any conclusion. It may also be used to see what happens if a negative number is chosen as a base. A discussion could then be pursued regarding why equations with bases having negative values are not included in our group of functions regarded as exponential functions.

REFERENCES FOR USING THE GRAPHING CALCULATOR

Best, G.W., and Penner, D.A. (1994). *Using the TI-82 to Explore Precalculus.* MathWare.

Cooney, T.J., and Hirsch, C.R. (1990). *Teaching and Learning Mathematics in the 1990s.* 1990 Yearbook. Reston, VA: National Council of Teachers of Mathematics.

Day, R. (February, 1996). "Classroom Technology: Tool for, or Focus of, Learning?" *The Mathematics Teacher.* Reston, Virginia: National Council of Teachers of Mathematics.

Fey, J.T., and Hirsch, C.R. (1992). *Calculators in Mathematics Education*. 1992 Yearbook. Reston, VA: National Council of Teachers of Mathematics.

Lund, C. (1995). *TI-82 Graphing Calculator Activities for Middle School Math*. MathWare.

Lund, C., and Andersen, E. (1995). *Graphing Calculator Activities Exploring Topics in Algebra I and Algebra II*. MathWare.

McDonald, J. (November, 1988). "Integrating Spreadsheets into the Mathematics Classroom." *Mathematics Teacher*. Reston, VA: National Council of Teachers of Mathematics.

Specht, J. (1996). *More Than Graphs: Activities for the TI Graphics Calculator*. Key Curriculum Press.

Williams, D.E., and Scott, T.L. (1993). *Investigating Mathematics with the TI-81*. Sunnyvale, CA: Stokes Publishing.

COMPUTERS

School systems, in general, have been slow to adopt the use of computers for a variety of reasons, the biggest of which is the tremendous amount of money needed to purchase them and then to purchase classroom sets of applicable software for all grades, levels, and subjects. Gradually computers found their way into our schools—first for administrative purposes, next in the school library, then in the business classes. Then followed computer literacy and programming. The next step was writing labs which made computers available for use by large numbers of students and teachers. Many schools now have one or more computer labs shared by all classes and subjects. Teachers continually look for good software, especially software usable in more than one subject.

In the mathematics classroom, we can use computers for computer-assisted instruction, linking the student directly to the material to be learned via the computer. It is used to enhance the education of students by presenting information, by guiding the learner, for practicing, and for assessing student learning. Computer-assisted learning includes drill and practice, tutorial, simulation, and problem solving. Other uses for teacher and students include programming, word processing, data storage and retrieval, modeling and demonstration.

Software designed for specific educational purposes usually focuses on basic skill development. In contrast, word processing, databases, spreadsheets, hypermedia applications, and multimedia can lead to a great many opportunities for problem solving and critical thinking. The Internet provides access for students and teachers to other countries, other classrooms, real math problems, research, and a way to exchange problems, challenges, and information with other schools.

Technology is most powerfully used as a new tool to support student inquiry, composition, collaboration, and communication. (Dwyer, *Teaching With Technology*) "Teaching and learning mathematics is not (and never has been) an observer sport." By doing mathematics and by becoming involved in mathematics, the students will learn more mathematics. As teachers we need to learn to be visionary and to be willing to try new teaching methods. We must formulate instructional goals, inquiry strategies and be able to guide student learning (*1990 NCTM Yearbook*).

USING SOFTWARE PACKAGES

When the computer arrives in your classroom, how do you use it for instruction without having a computer for each student and without having subject-specific software for them to use? This is the question many teachers face when a school system is determined to put a computer in every classroom but has not yet been able to purchase software.

You can use the graphing calculator on the computer to show graphs on the monitor or hook up the computer to an overhead display for instructional purposes. This can be used for many of the suggestions given in the graphing calculator examples, especially those for quick assessment and for the visual part of a lesson presentation.

For the middle school teacher the computer will take input and graph it in either pie graphs, bar graphs, line graphs, or scatter plots. This means that it can be used at many levels of instruction when doing graphs. Students can simply collect data and enter it without knowing how the graphs work and then gradually learn how to actually set up a graph on their own. This involves knowing percentages, ratios, computation, fractions, data comparison, pattern recognition, and using formulas or equations. The spreadsheet is also an instructional tool to use in the process of teaching equations. Real-life data related to topics the students are interested in can be used. This holds their interest and gives them the notion of the usefulness of mathematics.

TEXTBOOKS

Many of the recently published math textbooks have integrated technology into the text and/or have provided technological supplementary materials. *Geometry*, published by Addison-Wesley, is an example of a text developed to answer the needs of teachers wanting a student-oriented lesson designed to include technology. The teacher's text comes with an enrichment workbook, laboratory workbook, and a technology and lesson plan book that includes cooperative group activities. It also provides extra practice, tests, quizzes, reviews, and a test bank of questions on disk.

The Thomas' *Calculus* is another example of a text that integrates technology. It has graphing calculator exercises as well as references to computer software called *Calculus Toolkit*. This software is relatively inexpensive. It works well as a tutorial for students in addition to its usefulness for instruction. Many other Calculus texts provide the same type of technology integration.

The *Discovering Geometry* text is designed to be used with software called *Geometers Sketchpad*. This software is exceptional and can be used with most other high school math courses as well as geometry. *Discovering Geometry* also provides an appendix of "Logo Procedures for Computer Activities" which can be used on either an IBM computer or an Apple II without the Sketchpad software.

CALCULATOR-BASED LABORATORY

The calculator-based laboratory (CBL) provides students with opportunities for both hands-on activities and for interdisciplinary lessons. Students perform experiments using instruments such as the motion detector and probes for temperature, light, pH, and sound. These can be hooked up to the computer in the classroom and in turn to each student's graphing calculator for lab experiments. We use the Texas Instruments CBL. However, there are comparable versions available for other calculators. It records the readings on the graphing calculator. This enables the student to take individual results and, working with a group, interpret their data. This can be used in Algebra I through Calculus.

The following provide activities for the student and instructional guidance for the teacher: Using the TI-82 to Explore Precalculus by George W. Best and David A. Penner ; *CBL™ Explorations in Precalculus for the TI-82* from The Meridian Creative Group; *CBL™ Exploration in Calculus with the TI-82* from The Meridian Creative Group; and Vernier Software for the CBL.

CONCLUSION

"The instructional use of technology must reinforce students' learning. The emphasis should be on analytical problem-solving skills. The focus must be on concept development, not merely on procedures with limited use outside the classroom (e.g., computation). Computer technology must be fully available to all students and teachers at all times. Since computers are being integrated into every aspect of the workplace, the integration of technology into teaching and learning is essential. Teaching and learning require the careful sequencing of activities, not just 'topic after topic' sequencing in isolation with little benefit to students" (Swadener and Blubaugh, *1990 NCTM Yearbook*, p. 227).

6

Evaluation and Assessment in a Block Schedule

Teachers in a block schedule express concern about having enough grades for each student during a marking period. How many grades are enough? How often should a grade be taken? Should the types of grades differ from those collected when students are on a traditional schedule? Should the time for taking a test be extended to cover the entire block? Should students be able to use technology on the tests?

These represent many of the typical questions that arise when a school changes to a block schedule. The answers are as varied as the lessons that teachers design. What works for one teacher may not work for another. Some tests should be technology-active while others may need to be technology-inactive. Even though students meet a class fewer times during each course, there are still ample opportunities to "collect" grades. Teachers need to be clear about their purpose in taking a grade and convey that purpose to their students.

Any time teachers ask students to perform, they should indicate the purpose of the performance. A well-defined target is easier to hit than one that is constantly moving or even hidden from view. Students need to know what is expected of them and what degree of proficiency will be accepted. "Changes in curriculum and instruction must be accompanied by equally substantive changes in assessment policies, procedures, and instruments" (Stenmark, 1991, p. 7).

Also, students should be assessed using the same methods by which they originally learned the mathematics. It would be logical for teachers who integrate technology into their curriculum, to enhance the kinds of investigations and applications that students undertake, to allow this technology to be used for assessment purposes. Does it make sense to ask students to take measurements without using appropriate tools? Should a student be expected to carry out complex calculations without a calculator or computer if that was the method by which the calculations were done previously? The purpose of the assessment needs to make sense and the method by which students carry out the assessment task also needs to make sense. This chapter discusses the NCTM *Assessment Standards for School Mathematics*, the role of informal and formal assessment, the use of projects, and authentic assessment and rubrics. One chapter is by no means adequate for a complete elaboration of assessment techniques, but we offer as detailed a description as space allows so that teachers may pick and choose as well as get ideas for tasks that will work in their classrooms.

NCTM Assessment Standards

In a block schedule, both planning and instructional strategies must be adopted that encourage students' active engagement in the learning process. As teachers discover ways to plan and to organize activities for a block and design lessons that integrate technology, they must also develop meaningful forms of assessing student progress. "All students are capable of learning mathematics, and their learning can be assessed" (NCTM, 1995, p. 1). Also, "assessment should be a means of fostering growth toward high expectations" (NCTM, 1995, p. 1).

The Evaluation Standards propose that:

- student assessment be aligned with, and integral to, instruction;
- multiple sources of assessment information be used;
- assessment methods be appropriate for their purposes;
- all aspects of mathematical knowledge and its connections be assessed;
- instruction and curriculum be considered equally in judging the quality of a program (NCTM, 1995, pp. 1–2).

Many people use the "assessment" interchangeably with "evaluation." We adopt NCTM's definitions of both of these words: "Assessment is the process of gathering evidence about a student's knowledge of, ability to use, and disposition toward mathematics and of making inferences from that evidence for a variety of purposes" (NCTM, 1995, p. 3); and "evaluation refers to the process of determining the worth of, or assigning a value to, something on the basis of careful examination and judgment" (NCTM, 1995, p. 3).

INFORMAL ASSESSMENT

Our definition of informal tasks includes those activities that a teacher uses to get immediate feedback on students' understanding of a skill or a concept and in their understanding of it within the framework of the lesson, the unit, and/or its relationship to the study of mathematics. Teachers determine this by observing student behaviors as they walk around the classroom or work with students individually or in groups. Also, teachers gain information from student responses to questions, from student-initiated questions, and from small group and whole-class discussion.

Much information is revealed by a student's ability or willingness to attempt a task, by student discussion as they work on the task, through written and verbal attempts by the students to reach a solution, and by the questions that students pose as they work on the task. Watching students actively working on a task affords a teacher opportunities for monitoring and adjusting the lesson based on the needs of individual students and the class as a whole. This interaction between students and teacher improves the quality of assessments that can be made.

Depending on the purpose of a task, grades may or may not be assigned. Many times, teachers want to informally assess students to determine how things are going; to look for misunderstandings; to check if students understand a process, skill or concept; to decide if more time needs to be spent on practicing a process, skill, or concept; to find out if the class is together in its understanding; or to offer one more bit of practice before moving on. Assessing these tasks needs to be quick and painless, while obtaining maximum information. There are many ways to incorporate this within a traditional classroom or in a block schedule. Using varied methods provides a change of activity and keeps the students involved in both a traditional class and a blocked class.

DRY ERASE BOARDS

Several years ago, we bought a classroom set of dry erase boards and we had some large classes, only two panels of chalkboard, and next to no room to circulate around the room when the students were at their desks. It was easy to "cruise" the classroom by standing in one location while students held up their boards. We could see instantly who did or did not understand what we were doing and if the class-members were together in their mathematical understanding. Several positive side-effects emerged as a result of using the boards: we gained a method for getting all students to "practice" because they like writing on the boards; we found a way for students to risk doing an exercise or problem they might not normally attempt to do with pencil on paper because the boards are less threatening and they are not collected for grading; it is not wasting paper; and it is FUN!

A problem is written on the board or on an overhead transparency and the students work on it. Problems can be given for many purposes: as a warm-up activity, for homework review, or for guided practice. Problems can be as simple as solving a linear equation with one variable or as complex as graphing a set of data and then determining its equation. Students can work in pairs or groups to complete very complicated tasks and then transfer their work to their notebooks for future reference. They can experiment to discover patterns and "rules." Many of the ideas in the chapter on instructional strategies can be adapted for use with dry erase boards.

One word of warning: If your students have not played with these toys since elementary or middle school, they will want to draw at first. After they have gotten over the novelty of drawing, you can focus them for the real task and get surprisingly good results. Students never cease to surprise us in the detail they include in their dry erase work. Often it is more inclusive than that in their notebooks or on their tests. With effort, you can get them to transfer that same detail to other assignments.

Dry erase boards can be purchased from a school supply store or a more economical board can be made by purchasing tile board from a building supply store and having it cut into the size you want from four-foot by eight-foot sheets. Dry erase pens are not expensive, come in many colors, and will generally last for more than a school year even with active use. The erasers are washable and inexpensive. Washable overhead pens also can be used and then the boards may be cleaned with wipes used for babies.

USING THE OVERHEAD

Checking homework can be time-consuming and not beneficial to all class members. Asking students to volunteer to put solutions to particular problems on an overhead transparency and then to present that work to the class alleviates some of this wasted time. This affords teachers an informal way to check for understanding. It is also an opportunity to ask the class a follow-up question or two to clarify any misconceptions or to help students make connections to other concepts.

Students working in groups can be assigned a problem whose solution can be written on an overhead transparency and then presented by a member of the group to the rest of the class. Teachers can assign different problems to each group or have all groups work the same problem. A member of the group may either volunteer to present or a random selection can be made using the flip of a coin for students working in pairs, a spinner for groups of three or five, or a four-sided die for groups of four.

We favor the random approach so that students in the group work together to make certain that all members of the group understand the problem. If you use the volunteer method, it is helpful to limit and to record the number of times each person volunteers, so that all students have the opportunity to participate and some will not totally tune out and never participate.

The overhead projector also allows students to use the overhead graphing calculator. Students can exhibit data, show a graph, and discuss their findings. They can teach other class members how to enter information on the calculator, freeing you to observe and to help others in the class.

SPAGHETTI GRAPHING

The first time we saw a teacher use spaghetti graphing was at an NCTM National Conference. She had connected a computer to an overhead palette on an overhead projector and was running Sunburst's *Green Globs and Graphing Equations*. She entered linear equations on the computer and then had the "students" graph them on coordinate grids using uncooked spaghetti for their lines. It was a very effective way to determine who needed more help, because it freed the teacher to observe students at work—not to mention that it was lots of fun!

For spaghetti graphing, students need a coordinate plane on notebook-size paper and some uncooked spaghetti. (Students are prone to chew on the spaghetti so they need several pieces. A nonconsumable alternative is wooden shish kabob skewers.) You can copy a coordinate plane on heavy cardstock and laminate it to last. Also, if skewers or spaghetti are unavailable, students can use dry erase pens to draw the lines and then the grids can be wiped off.

This computer program is an excellent way to integrate technology but is not necessary to this technique. A linear equation can be written on the overhead or on the chalkboard to accomplish the same purpose. In this instance, use of the technology is not being assessed. The student's ability to graph a given linear equation in two variables is what is important.

A variation of this can be used for graphing quadratic equations and other curves. It takes a bit of time and effort but the benefits are worthwhile. We bought very small straight pasta tubes at the grocery store and threaded them together on fine fishing filament to make a movable "necklace" about 10 to 12 inches in length. This string can be placed on the coordinate grid and shaped to model the graphs of nonlinear functions. Also, packages of "Wikki Stix" (flexible, waxed candlewicks about 8 inches long) may be purchased at toys stores. The wicks stick either to the dry erase boards or to laminated grids, may be twisted to form any curve, and last.

JOURNALS

We have discussed using writing and journals in the chapter on instructional strategies, but a few words are merited here because journal writing can be assessed informally without assigning a grade to each entry. Asking students to respond to a question, to explain a process, or to summarize material affords teachers a powerful instrument for determining student understanding. Journal entries may be made at any time during the class when they are appropriately tied to the instructional process.

On a personal note, if teachers assign journal writing, they must read the entries of their students and respond to them or students will not think that the teacher feels the experience is really worthwhile. An example of the nonvalued writing exercise is having students write copious, detailed summaries of chapters each grading period only to count the number of pages in each summary and assign a grade based on quantity of pages. The teacher returns the work, unread, with a grade and a scribbled note across the summary attesting to its quality. Students quickly pick up on this and begin substituting "garbage" between the first and last pages just to make the pages add up. The moral of the story: if you want students to benefit from a writing experience and you do not want "filler," read and respond to the writing—or it is probably best not to make the assignment.

If students write many entries, tell them you are not going to be able to read all of them if you do not intend to. Be honest and up front. Let them know that you will pick some and respond to them. Ask students to tell you if there are particular entries for which they would like your response, or if there are any that they would like to rewrite or for you to skip. Writing is an excellent method for getting to know your students, not just their mathematical strengths and weaknesses.

One of the first assignments we ask our students to complete is a mathematical autobiography, from their first memory of numbers to their most recent. In their writing, we want them to include their favorite mathematics' memory as well as their worst mathematics' nightmare. We ask them to describe their ideal mathematics class. What would the class look like? How would you like to learn mathematics? What characteristics would your ideal teacher have? And, finally, we ask , tell us what you want to be able to do mathematically when you finish this course. We respond to their writing by the next class. It is not always lengthy, but we include comments that let each student know we read what was written. We make notes about what each student hopes to get out of the class and ask them for a follow-up response at the end of the course. This helps us assess the students' mathematical disposition for doing mathematics, gives us insight into their preferred learning styles, and lets us get to know them as individuals.

In the beginning, when students are first starting to write, we try to make some written response to each student. As time passes, their quality of writing improves along with their mathematical understanding and power, so we can read and respond less.

If the assignment entails explaining a process, students may peer review each other's works to determine if the writing is clear before being assigned an entry to be formally assessed. They should be able to read the other student's work, follow their logic step-by-step, and reach the same result as the writer. If the writing is unclear, students make suggestions to each other on how to improve the entry and then make corrections. This is a particularly good method to use with explanations on how to solve equations such as $3x + 7 = 7x - 11$ in which each step must be justified.

FORMAL ASSESSMENT

"The assessment process can be thought of as four interrelated phases that highlight principal points at which critical decisions need to be made" (NCTM, 1995, p. 4). They are: planning the assessment, gathering evidence, interpreting the evidence, and using the results. These phases are neither linear nor sequential. They are guides in the decision-making process. Assessment is a complex process in which teachers should know the purposes for doing assessments, how the decisions will be made after the assessments are done, and the standards to which the assessments are held. (NCTM, 1995, p. 5)

The Assessment Standards for School Mathematics includes "six assessment standards which are the criteria…for judging assessment practices" (NCTM, 1995, p. 5). The standards are:

- Assessment should reflect the mathematics that all students need to know and to be able to do.
- Assessment should enhance mathematics learning.
- Assessment should promote equity.
- Assessment should be an open process.
- Assessment should promote valid inferences about mathematics.
- Assessment should be a coherent process. (NCTM, 1995, pp. 11–20)

As we discuss formal assessment, we will keep these six standards in mind.

Formal assessment includes all of the traditional pencil and paper quizzes and tests and also incorporates alternate assessment grades from projects, learning logs, long-term problem-solving activities, and authentic assessment. However, the traditional testing instruments should change to include more open-ended questions and authentic tasks. Therefore, there should be fewer items on each test. We discuss authentic or performance assessment in a separate section of this chapter and briefly discuss portfolio assessment, identifying resources that provide a more complete discussion for implementing it into the classroom.

Students should be evaluated using the same methods and processes by which they were instructed. If they used calculators in their classroom exercise, they should be allowed to use calculators on those parts of the graded work that are designed for calculators. If students work in groups during class time, then opportunities for working together for part of the test might be created as an option on some quizzes and tests. This way assessment integrates all of the methods by which students acquired their mathematical knowledge.

QUIZZES

PARTNER QUIZZES

One of the positive features of a block schedule is having the time to let students work together to develop their mathematical understanding. If students have cooperatively worked on problems together during class, then it makes sense to give them opportunities within the course to work together on a set of quiz problems.

On a partner quiz, each pair of students is given three copies of the quiz: one quiz for each to use for individual work and the third to be turned in for a grade. The students are given 10 to 15 minutes to work alone on the problems. They are then given an additional 10 minutes in which they compare answers and complete the third copy to turn in for a grade. If partners differ on answers, they must both rework the problem together, looking for an answer upon which they both agree.

Another method allows students to work in groups of four or five. They work alone and then confer on their solutions. All members of each group are encouraged to pay careful attention to the discussion when they confer and make corrections. The teacher can then go around to each group and randomly choose one paper from each group to grade. Randomly choosing one paper really encourages group cooperation and listening skills.

At the beginning of a course, we let the students choose their own partners. Students who wish to work alone are encouraged to work with a partner so they realize that mathematics is not always one person working alone in isolation. This is particularly beneficial for applications and problem-solving activities. However, we do not force students to work together. Also, if we feel that the partnership is lopsided and one student contributes most of the time while the other student merely copies most of the time, we "arrange" for an apparent random partner assignment the next time. This takes the pressure off the donor and places the responsibility for learning mathematics squarely on the shoulders of the recipient.

Working alone at first and then together stimulates positive mathematical discussion. Students who do not feel comfortable in whole-class interaction usually have something to contribute to the partnership discussion and must contribute if answers differ.

Some teachers are not comfortable with partner work because of a fear that one student does all of the work. Not every quiz is taken with a partner or group. Students need to be assessed individually as well. This method is only one of multiple ways to assess student understanding and progress. It is only one part of the picture of a student's mathematical ability and achievement. The purpose of working with partners is to enrich the mathematical experience and to help build confidence in students' abilities to do mathematics.

ON-YOUR-OWN QUIZZES

This type of quiz is just what its title implies: a student works alone and turns in the completed work. Quizzes can be anywhere from 5 to 30 minutes in length and can be used to evaluate individual skills, concepts, and/or processes. They may reflect information from one aspect of a unit or from several lessons in the unit.

TESTS

ON-YOUR-OWN TESTS

An on-your-own test is designed for the students to work by themselves. Depending upon the skills, concepts, and processes being evaluated, it may be either calculator-inactive or -active for part or all of the test. Normally, these tests take about 60 minutes.

When we create the test, we incorporate a variety of questions to assess skills, concepts, and processes. Questions include multiple choice, quantitative comparisons, open-ended short-answer questions, discussion questions, process problems, and applications. Students need to communicate mathematically as well as to find solutions and make calculations.

Discussion questions that ask, "Is it possible...?", "How would you explain this to... (choose a target audience that would not easily understand a concept)?", or "Given this solution...explain, specifically, where the reasoning is faulty and correct the solution," are mathematically rich in the responses they should elicit. Some examples of these are given in Appendix A, which contains lessons, activities, and assessments.

Students are not naturally proficient in writing, especially in mathematics classes. Their responses should be written in complete sentences using appropriate mathematical vocabulary. They need to see examples of the quality that you expect from them. By classroom modeling and assigned practice, you can help students become proficient at writing responses to discussion questions. Care should be taken to word the question so that both you and the students know the type of response that should be given. This does not mean that you should provide students with the answers you want, but that you should work with them so they understand the types of information that should be included in the response to a mathematical discussion question. Using peer review also improves the quality of communication. The English department will appreciate your efforts as well.

PARTNER TESTS

Partner tests allow two or more students to work together during one part of the test. We usually tailor partner tests so that students work together during the first 30 minutes of the test on problems that involve those activities they did together in class. We use the beginning of the test as a review and to build mathematical confidence in the students.

Usually, the partner part either consists of problem solving or data analysis of a real-world application, or a few well-designed application problems that reinforce the concepts in the unit. Generally, the partner part of a test counts as 25 percent of the test grade. This is not weighted so heavily that a student who is not prepared for the individual part does well overall.

Linear programming problems, exponential growth and decay applications, and data analysis are some of the topics that lend themselves to partner work. Some examples of these are included in Appendix A, which contains lessons, activities, and assessments.

TAKE-HOME TESTS

When students take a test home, many times red flags go up from other teachers and from parents. There is concern that students will use unacceptable resources for completing the test or that the test is much harder than an in-class test. A teacher needs to let students know why the test is to be completed out of class and what the teacher expects of them. Students need to know what resources they may use while they work toward solutions and whether they can talk to others about the problems. Most students will usually respond positively to expectations that are clear and equitable.

When we consider giving a take-home test, we ask ourselves what we want to assess. What is the purpose of having students take the test home? Is this a test which can be completed better out of class than in class? What types of tasks do we expect our students to do? Will this test build the students' mathematical confidence and increase their mathematical power? Are we saving class time because our pacing is off or because students have missed many days due to inclement weather or some disaster?

If the answer to the last question is a resounding yes, then we need to rethink our planning so that we can make the most of the time we have left for the class.

There may be concepts that can be woven together to ensure mathematical understanding. If we are running out of time, more than likely the students will not be prepared for a test that evaluates more than skills.

As a rule, we do not use take-home tests to assess mathematics at the skill level. Instead, students need to be able to apply the skill in a problem-solving situation in which they may be asked to work with real-world data on a take-home test. We view a take-home test as an opportunity for students to reflect upon their base of mathematical knowledge and to choose appropriate skills, concepts, and processes to solve problems. We want them to think about what they are doing and to try different approaches to the problems, not feel the pressure of a time limit. We want them to treat their responses as they should treat an English composition: each should include an outline, a rough draft with revisions, and a final draft. An out-of-class test should challenge students metacognitively; it should encourage them think about their own thinking.

ALTERNATIVE ASSESSMENT

We define alternative assessment as those tasks or situations that differ from the more traditional assessment quizzes and tests. While some of the previously discussed assessment types may be considered somewhat alternative in nature because of their use of open-ended questions and problem-solving orientation, we exclude them from this discussion.

LEARNING LOGS

Learning logs are journals that help students make connections between verbal and mathematical literacy. Entries should be carefully planned to elicit the best of the student's thinking. They may include verbal descriptions, pictures, diagrams, tables, and any other student materials. Responses should be in complete sentences using appropriate mathematical vocabulary and should include any of the ideas and work that the student had as he or she was writing the entry. The log entries may be kept in a separate section of the student's notebook.

In the beginning, model the process with the class as a whole. A question is posed to the entire class and students are asked to think about the question and to write down a short response. (This is one way to use dry erase boards.) A possible question for a Prealgebra or an Algebra I class might be, "How could you determine the slope of a handicap ramp?" After the students have had a chance to jot down their ideas, the teacher asks students to give some of their responses, which are written on the chalkboard or overhead. Next, the teacher models a response for the students, carefully discussing which elements are used to make an acceptable response. The students have the teacher's response as a model for their writing. Students should be encouraged to write a draft and then to make revisions so that their final response expresses clear mathematical understanding. Teachers should repeat this process over several weeks so that students have enough practice to formulate competent responses of their own. They will realize that writing is a valuable experience.

Learning logs can be used **before, during**, or **after** a lesson. A "before" log elicits responses that serve as advance organizers to assess prior knowledge of a concept. A "during" log asks students to describe their thinking processes or to explain the use of a procedure. An "after" log gets students to summarize a lesson or to evaluate their success in using a problem-solving method.

Using learning logs regularly "promotes reflection and self-assessment by students" as well as "provides teachers with a sequential record of their students' thinking and conceptual learning" (NCTM, 1993, p. 96). Teachers who use portfolio assessment may have students select responses for inclusion in their student portfolio as examples of their mathematical thinking and literacy.

PORTFOLIOS

A mathematical portfolio is a collection of a student's work that includes "many types of assignments, projects, reports, and writings. Progress in, attitudes toward, and understanding of mathematics can be seen in a comprehensive way" (Stenmark, 1991, p. 35). Students choose the work they want included in the portfolio. It may be their best work from a given set, or selections may show improvement from earlier selections. Students confer with their teacher about portfolio choices and about mathematical growth. The teacher and student gain a more complete picture of the student's progress than with only tests and quizzes.

NCTM supports the use of portfolios because they demonstrate that multiple sources of information provide a much better picture of a student's mathematical understanding and growth. Portfolios are ongoing works in progress. Analyzing them at regular intervals allows teachers to make course adjustments.

Students should be responsible for maintaining their portfolios and keeping them current and organized so that they show the work that best represents them. Students should know the purpose for creating the portfolios. Are they to be used for parent-teacher conferences, or for evaluation by someone outside their classroom, or for their own self-evaluation and self-reflection? The portfolios need to be accessible to the students and the teachers on an ongoing basis in the classroom. Portfolios become part of a learning process, not just a nice place to keep work.

Mathematics Assessment: Myths, Models, Good Questions, and Practical Suggestions (1991), edited by Jean Kerr Stenmark, examines mathematical portfolio assessment in detail. The discussion includes "inside a sample portfolio, selected goals and suggestions for contents, managing student portfolios, assessing portfolios, and writing in mathematics" (Stenmark, 1991, p. 36). The focus described for student portfolios is on these items: "student thinking, growth over time, mathematical connections, student views of themselves as mathematicians, and the problem-solving process" (Stenmark, 1991, p. 37).

The use of portfolio assessment is growing. Many teachers use it in conjunction with traditional tests and quizzes to develop a better picture of student mathematical understanding. Students like the portfolios because they give them the opportunity to practice and edit. Also, they can see their growth as time passes.

We have not used portfolio assessment in its entirety, but have begun slowly to build pieces of it into the learning process. Our students keep a comprehensive notebook of class notes, homework, problem solving, handouts and practice exercises, writing exercises, quizzes and tests. We have them pick and choose items from these for us to assess as a whole during each marking period. Each year we try to move closer to true portfolio assessment to get a better picture of student learning and growth.

We encourage you to consult some of the references we have listed for a more complete description of using portfolio assessment. *Mathematical Assessment* is an excellent starting place because of its detailed descriptions and examples.

PROBLEM SOLVING

"Answer-focused paper-and-pencil tests are by far the most common type of assessment of mathematical problem-solving progress" (Charles et al., 1987, p. 4). However, well-designed multiple-choice and open-ended questions may be used for paper-and-pencil test problem solving. Students' problem-solving progress also may be evaluated through observation and questioning, interviews, and self-assessment. Charles, Lester, and O'Daffer present a comprehensive discussion of problem solving in *How to Evaluate Progress in Problem Solving.*

Assessing problem solving should be based on how you teach problem solving. What are your goals for including problem solving in your curriculum? Charles, Lester, and O'Daffer identify seven goals for teaching problem solving:

- "To develop students' *thinking skills*
- "To develop students' abilities to *select and use problem-solving strategies*
- "To develop *helpful attitudes and beliefs* about problem solving
- "To develop students' abilities to use related knowledge
- "To develop students' abilities to *monitor and evaluate their thinking* and progress while solving problems

- "To develop students' abilities to *solve problems in cooperative learning situations*
- "To develop students' abilities to *find correct answers* to a variety of types of problems" (Charles et al., 1987, p. 7).

Before discussing methods for evaluating problem solving, we should consider what types of questions lend themselves to the process. Problems should be rich in mathematical content and should lend themselves to inquiry and discovery. They should engage and involve students in doing mathematics. Students should be able to approach the situation using any of a number of heuristics (working backwards, trial and error, drawing a diagram, making a table of values, etc.) to reach a reasonable solution. Problem solving is more than applying skills, concepts, and procedures. "It is a process by which the fabric of mathematics...is constructed and reinforced" (NCTM, 1989, p. 137).

Many times, the problems chosen for investigation are nonroutine in nature. For example, how many different ways are there to make change for a 50-cent piece? Students need to develop strategies for keeping track of their "ways" and at the same time be able to explain how they decided to approach the problem and how they arrived at their solution. While this problem may seem rather easy, for students who need to develop organizational skills it is a complex problem.

Other problems may be directly tied to the curriculum. Students might be asked to determine if the handicap ramps within their school or some other building are within the community's building code. This would require them to find out what the regulation is for the ramps, take measurements of the given ramps, determine the slopes of the ramps, and then write an explanation of their findings.

Several years ago, we developed a biweekly set of problems in which the students were to actively engage in solving problems. Many teachers call these problems of the week (POWs) and assign them on a weekly basis. We discovered that on our alternating block, two weeks worked as a better time frame for students. We named them CHOMPS for "Challenge: Higher Order Mathematical Problem Solving," and adapted the write-up model from the Interactive Mathematics Project (IMP). We must encourage students to put their thoughts and findings into words. Communicating mathematical ideas is one way to empower students mathematically and to take a bite out of mathematical illiteracy.

We provide the students with a handout on general rules for presenting their solutions and make sure they understand exactly what they are to do for each problem. We model several examples for them to use in their initial efforts. If they know the expectations, then they can meet or exceed them.

In the write-up, the students must explain the problem in their own words; discuss their thinking while they were working toward a solution for the problem, including methods that did not lead to a successful solution; state the solution or partial solution, justifying why it is the best answer; write any questions that might arise as a result of this investigation or that might be an extension of it; and, finally, evaluate the difficulty of the problem with an explanation (NCTM, 1993, pp. 213–14). Each part of the problem is assigned 0, 1, or 2 points. Zero points if the part is missing, 1 point if there are flaws or misconceptions in the response, and 2 points if the answer meets the expectations.

Modeling the process for students takes time; however, it is a critical part of instruction. Working as a class to solve a problem, provides students with a beneficial experience that enhances their ability to do mathematics.

There are many good sources for problem-solving tasks: the *Mathematics Teacher* monthly calendar, Theoni Pappas' books, Martin Gardner's books, and others listed in Appendix B: Resources. *Problem Solving Strategies: Crossing the River with Dogs and Other Mathematical Adventures* is an excellent textbook for teaching problem solving.

PROJECTS

Projects can address a unit, they can be topic- or concept-specific, they can serve as an advance organizer to assess students' prior knowledge, or they can serve as culminating activities at the end of any given time period. Projects can be short-term or long-term. They can be used to assess the student's knowledge of concepts and ability to apply his or her mathematical learning to a new situation.

When studying linear equations, students could be assigned to make a unit-culminating poster illustrating the use of a linear equation in the real world. For example, one of our students who swam for a local swim team was interested in the finish times of women in the 200-meter butterfly. She did the research and presented her findings on a poster complete with a graph showing that the finish times over the years diminished almost linearly. She analyzed the finish times to conjecture what the absolute fastest time could ever be. She hypothesized that sooner or later, the finish times would level off and not decrease further because there has to be a limit to the body's tolerance and speed. She was able to apply the mathematics she had learned to a situation in which she had great interest and involvement. The poster was graded on her data representation and analysis, her correct use of mathematical concepts, her conjectures and conclusions, her poster's attractiveness, and her presentation to the class.

If you assign projects, you should model some examples of those that are acceptable and those that are unacceptable. We usually use projects from previous years to illustrate good examples and we create "bad" examples ourselves so that students who performed poorly are not singled out.

AUTHENTIC ASSESSMENT AND RUBRICS

> Authentic Assessment tasks highlight the usefulness of mathematical thinking and bridge the gap between school and real mathematics. They involve finding patterns, checking generalizations, making models, arguing, simplifying, and extending—processes that resemble the activities of mathematicians or the application of mathematics to everyday life. When we see students planning, modeling, and using mathematics to carry out investigations, we can make valid judgments about their achievement. (Stenmark, 1991, p. 3).

Designing authentic tasks takes time because they must be well-written, worthwhile experiences in which students want to become involved. However, the amount of information a teacher gains from the student's response makes the effort valuable. Rubrics (descriptions of what standards will be applied in evaluating a task) need to be specific and clear for each open-ended assessment task. A general rubric was given in the problem-solving section, but there are many others that may be applied.

Teachers working together are able to design authentic assessment tasks more easily than those working alone. Each teacher brings ideas and insights to the process and good assessments are more efficiently designed. Also, reading responses together ensures consistency in evaluation, especially in the beginning when benchmark papers (exemplars) need to be determined.

In Appendix A, which has lessons, activities, and assessments, there are several authentic assessment tasks complete with the rubrics we use in our county. The Cake Task was designed using information from *Guinness Book of World Records.* It is an excellent resource for designing authentic assessments, especially for evaluating students' ability to work with very large or very small numbers, to express an understanding of quantities, or to compare quantities. The Recycle Problem was adapted from the California Assessment Program.

CONCLUSION

In assessing student progress and achievement, we need to ask ourselves these questions:

- What do we want students to be able to do mathematically?
- What should they be able to use to do mathematics?
- How will they show us that they know how to do mathematics?
- How will we determine if they know how to do mathematics?

We must be able to answer each of these questions for students to have a clear understanding of what is expected of them. If we define the target and keep it from moving, students will be able to hit it more consistently.

APPENDIX A

SAMPLE LESSONS, LABS, AND ASSESSMENTS

RECOGNIZING POLYGONAL SHAPES

Subject: Pre-Algebra, Middle School

Objectives: To recognize basic polygonal shapes. To formalize the concepts and formulas for perimeter and circumference. To practice the application of perimeter and circumference formulas. To share specific life examples of perimeter. To determine the perimeter of a polygon and the circumference of a circle.

Materials: A set of tangrams for each pair of students. A classroom set of MIRAs. Measuring tapes. Rulers. Construction paper. Collection of round tins, cans, pie plates. *Pre-Supposer* computer software.

Activities During the Lesson: (This lesson takes two blocks.)

1. Use tangrams to make squares.
2. Groups write their own definitions of various polygons and share them.
3. Students measure perimeter of tangram pieces and look for shortcuts or formulas in determining them.
4. Students measure circumference of round objects and use the MIRA to find the diameter for each.
5. *Pre-Algebra With Pizzazz* practice CC 27, 28, 29, and 31.
6. *Pre-Supposer* computer activities.

Warm-Up Activity: Use the tangrams as the warm-up in this lesson.

Lesson Outline:

- Using a set of tangrams, student pairs will make as many different squares as possible. Each square will be traced (including the interior shapes) on a piece of construction paper.
- Class development, through discussion, of a list of the names and primary characteristics of a triangle, square and rectangle (emphasize those with equal sides or pairs of sides).
- Teacher-directed development of the formulas for the determination of the perimeter of a rectangle, square, and triangle.
- Students will measure the tangram pieces and square tracings and find the perimeters and also practice using their formulas.
- Continued discussion of the polygons of the tangram, including the parallelogram, pentagon, and rhombus, and the associated formulas for the determination of perimeter.

- Write in the student journal—Describe how at least three of the polygons discussed so far are used in your home (house construction, home furnishings, or interior or exterior decorations).
- Class discussion of the primary characteristics of a circle (radius, diameter, center).
- Working in groups of three, students will measure the circumference and diameter of several round objects using a measuring tape for the circumference and the MIRAs and a ruler for the diameter. These are to be recorded on a chart or table having three columns, one for the circumference, the second for the corresponding diameter, and the third for the ratio of the two (circumference divided by the diameter).
- Using the collected data, the class will develop the formula for the circumference of a circle ($C = 2\pi r$ or $C = \pi d$) and examine the role of π.
- Computer lab using the *Pre-Supposer* software.
- Student practice.

Lesson Extension or Activity: Students are to select and complete one of the following:

- A sketch of a room in your home, including its furniture. Then determine the perimeter of the room and that of at least three objects (such as a chair) that sit on the floor.
- A sketch of your yard including flower beds, sidewalks, driveways, and the house. Then determine the perimeter of the yard and of at least three other areas in the yard (such as a flower bed).
- A sketch of an athletic field (baseball, football, soccer, or hockey field, or basketball court) including all field or court markings. Then determine the perimeter of the field and of at least three other parts of the field (such as the end zone).

FINDING THE MEAN, MEDIAN, AND MODE FOR DATA

Subject: Pre-Algebra or Algebra 1, Middle School

Objective: Students will be able to gather data and use it to find the mean, median, and mode for the data. They will represent their findings in a graph using percentages and fractions. Students will also observe data, look for patterns, and form conclusions based on newly found information.

Materials: Individual bags or boxes of M&M candies, paper pie plates, small paper cups, protractors, straight edge, preprinted tables or charts, calculators.

Prior Assignment: Students are to find newspaper or magazine articles and advertisements, or television ads that use statistics in some way. They are to bring the actual ads or copies of the ads to class.

Activities During the Lesson: (This lesson is designed for one block of time of approximately ninety to one hundred minutes.)

1. Groups of three will share the advertisements.

2. Individually the students will sort and count the candies. Record the information on a table provided by the teacher.

3. Form groups, share data and complete table.

4. Use the computer to make a pie chart or, if a computer lab is not available or your students need practice using percentages and a protractor, they can make their own pie chart.

Warm-Up Activity:

1. Using prerecorded commercials that use percentages or ratios in the ad, ask students to write down what inference is being made; or

2. Select an article from the newspaper in which percentages are used to persuade the reading audience to share the same view as the writer. Students are to respond by stating whether or not they are convinced and why or why not.

Lesson Outline:

- Students share their advertisements with the class noting the percentages and statistics used in the ad. The teacher uses these ads to introduce the concept of collecting data and forming statistical information.

- Each student is given a package of M&M candies, a paper plate, protractor, straight edge, and a small paper cup to hold the candies. Students use the paper plate to contain the candies while sorting and counting.

- Students count the number of M&M candies in their box and record the result. They also sort the candies according to color and count the number of candies in each color and record the data. These are recorded on a chart provided to all students.
- Teacher-directed development of the definitions of the terms mean, median, and mode.
- Students form groups of three to five depending on the size of the class. These groups use the data they have collected and find the average number of candies in a bag and the average (mean) number o f candies of each color in a bag.
- Upon completion of the table for each group a representative from each group places their data together on a table on the overhead or on the blackboard.
- Students are asked to find the following using the numbers from the entire class: average number of candies in a bag; the average number of each color of candy in each bag; the median number of candies in an individual bag; median number of each color; the mode for the number of candies in an individual bag; the range for the number of candies in a single bag.
- Teacher-directed information on making a pie chart either by using a protractor and percentages or by using a computer and appropriate software.
- Using the collected data students make a pie chart to show the percent of the candies in a bag for each color. This is compared to the information published by the Mars Candy Company at its Web site.
- Student practice.

Lesson Extension or Activity: Students are to design their own activity/project to include collecting data, finding mean, median, and mode, making a pie chart and then presenting their finding to the class.

COLLECTING, GRAPHING, ANALYZING, AND INTERPRETING DATA

Lesson Title: Something's Afoot

Subject: Pre-Algebra or Algebra I

Objectives:

- To collect, graph, analyze, and interpret data
- To create a linear model from scattergram data which is linear
- To develop a mathematical argument supported mathematically and presented in a well-written, paragraph format

Materials:

- Data collection and lab sheets
- Yardsticks, tape measures, or meter sticks with English units
- Pencils, graph paper, and notebook paper

Lesson Activities: Work in pairs to:

- Collect data by taking measurements
- Write individual data on on overhead transparency
- Compile classroom data on the data collection sheet
- Answer lab questions

Individually:

- Write the report using the data collected

Time Required: 85–90 minutes (The report can be written as homework.)

Warm-Up Activity: Give students several sets of points and have them write the lines in slope intercept form or show them some lines and have them find their equations in slope-intercept form.

Lesson Outline:

- Explain the activity to the students and give them the worksheets and materials they will need.
- Have the students collect the data in pairs, record their information on the transparency, and then copy the results of the class to their data collection sheet.

- Circulate as they work to make sure that students are following directions when measuring, filling out the data collection sheets, and answering the lab questions.
- Have students compare their individual equations to make sure that when they leave class, they can write their report.

Extension/Culminating Activity: Have students use a graphing calculator to graph their data points and then use the linear regression function to find a line of best fit. Have them compare their approximate line to the calculator's and discuss why they might be different.

Something's Afoot!!

Some people say a person's shoe size is directly proportional to his or her height. Tall people have large feet and short people have small feet. Other people cite exceptions of tall people with small feet and short people with large feet. This group believes that a comparison between the length of a person's foot and the length of his or her forearm from wrist to elbow describes a more direct relationship than height to shoe size.

In order to determine which of these groups poses the more valid argument, you are to conduct a survey in which you collect data about your classmates and then organize, graph, and interpret that data.

Lab Instructions:

With a partner, determine your height **(in inches)**, your shoe size (women will need to subtract 1.5 shoe sizes from your women's size to get an equivalent male size), the length of your foot from middle toe to heel **(in inches)**, and the length of your forearm from your wrist to your elbow **(in inches)** using either a yardstick/meter stick or a tape measure.

On the overhead transparency, record each of your pieces of information and then copy all the material from the transparency onto your own data collection sheet.

After you have completed filling your data collection sheet, sketch two separate graphs to represent your data: on one graph, plot height as the independent variable **(x)** and shoe size as the dependent variable **(y)**; on the second graph, plot the length of each foot as the independent variable **(x)** and the length of each forearm as the dependent variable **(y)**.

After you have plotted all of your data points, follow the rest of the lab sheet to arrive at your own opinion about the foot controversy. You must argue for or against one or both of the positions using the data you have collected, graphed, and analyzed. Your decision should be presented in a well-written paragraph or two, using valid mathematical arguments confirmed by your data and analysis.

Something's Afoot
Data Collection Sheet

	HEIGHT	SHOE SIZE	FOOT LENGTH	FOREARM LENGTH
1				
2				
3				
4				
5				
6				
7				
8				
9				
10				
11				
12				
13				
14				
15				
16				
17				
18				
19				
20				
21				
22				
23				
24				
25				
26				
27				
28				
29				

Lab Activities and Questions:

1. Do most of the points on each graph seem to come close to falling on a line? **(One way to tell is to draw as narrow an oval as possible around the points. The narrower the oval, the closer the points are to lining up. The more circular the oval, the less so.)**
2. Draw a line through the middle of each scatter plot. The line should also run through the middle of each of your ovals.
3. Your lines are **geometric models** for the relationship between height and shoe size and between foot length and forearm length.

   You can use your first model to estimate your shoe size. Find your height on the horizontal scale and move upward until you come to the line you have drawn and move across to the vertical scale to read your predicted shoe size.

   What shoe size was predicted? ____________________

   What is your actual shoe size? ____________________

   How does this compare to your actual shoe size? ____________________

   You can use your second model to estimate the length of your forearm. Find your foot length on the horizontal scale and move upward until you come to the line you have drawn and move across to the vertical scale to read your predicted forearm length.

   What forearm length was predicted? ____________________

   What is your actual forearm length? ____________________

   How does this compare to your actual forearm length? ____________________

   Which is these models do you think is better for estimation? Why?

   ____

   ____

   ____

4. Choose two points on your first line. Try to make them integers if possible. They do not have to be points of your original data but should be fairly far apart.

   Line 1: Height vs. Shoe Size (_____ , _____) and (_____ , _____)

   Do the same for your second line.

   Line 2: Foot Length vs. Forearm Length (_____ , _____) and (_____ , _____)

5. Use each of the sets of two points to find the slope and y-intercept of each line.

Slope of Line 1: ________________ y-intercept: ________________

Slope of Line 2: ________________ y-intercept: ________________

Line 1 Equation: ____

Line 2 Equation: ____

6. These equations are **algebraic models** for your relationships.

   Use model one to calculate your predicted shoe size by substituting your height, **x**, into equation 1 and rounding the shoe size to the nearest 0.5.

   What shoe size was predicted? ________________

   How close is the predicted shoe size to your real shoe size? ________________

   Use model two to calculate the predicted length of your forearm by substituting your foot length, **x**, into equation 2.

   What forearm length was predicted? ________________

   How close is this to your actual forearm length? ________________

7. Using a graphing calculator, enter the data from each of the above models and have the calculator determine the line of best fit.

   Which of the models, one or two, is a better predictor? Why?

8. Write your report on a separate piece of paper, using your analyses to defend or refute one or both models.

PIECE OF CAKE

Title: Piece of Cake

Grade/subject: 7–10th grade mathematics

Description: Part 1: The student is asked to calculate the cost of making the world's largest cake by determining the amount of each ingredient needed.

Part 2: The student is asked to estimate the amount of trash generated by weighing the actual materials and computing the cost per pound for the cleanup needed to make a $1000 profit.

Time Required: Part 1 = 45 min; Part 2 = 45 min

Materials: Part 1: Students need paper, pencil, and calculators.

(Optional: teacher may supply students with prices from various stores or may require students to do comparison shopping.)

Part 2: If the teacher wants each student to do the work, then each needs an egg carton, an egg shell, an empty oil bottle, an empty cake mix box, an empty plastic cake bag, and an empty icing container. The teacher can have the class do the weighing as a group and then have students work individually.

(Optional: teacher may require students to find out what the recycling and landfill requirements are for their particular locality.)

Objectives:

- Systematically organize, describe, and, when appropriate, collect data
- Apply and solve ratios in proportions
- Make inferences and convincing arguments that are based on data analysis

Outline:

Part 1: Students are asked to calculate the cost of making the world's largest cake. They must determine the amount of each ingredient they will need to purchase. Their answers should show calculations for determining the amounts of each ingredient to be bought and the correct calculations for the costs of the ingredients. They should provide a justification for their choices. Their final answer should include 4.5% sales tax.

Part 2: Students are asked to estimate the amount of trash generated from the creation of this cake by weighing actual materials and computing the cost per pound for cleanup. Their goal is to make $1000 profit. Their answers should show their calculations for the amounts of trash generated, both recyclable and non-recyclable. Their proposal should be written in clear understandable language with their reasoning for their bid.

What a Piece of Cake!

> The largest box-mix cake ever created was a 90,000-lb 8,800-sq-ft "Duncan Hines" cake that used 31,026 boxes of a yellow cake mix, 20,000 layers of 3 lb each, topped with 30,000 lb of vanilla icing in 1989. It took 24 hours to assemble...after 32 hours of baking....The cake was big enough to serve 300,000 people."
>
> *Guinness Book Of World Records*

Task One:

You are in charge of purchasing all of the ingredients for making this monstrous cake. Using the prices from the table below, decide which brand you will use. Write up your results showing your work and explaining your **mathematical** reasoning for the decisions you made. What is the total cost for making this cake, including 4.5% sales tax on you purchases? (Note that water is free.)

Remember:

1 cup = 8 oz

1 gallon = 128 oz = 16 cups

4 sticks = 1 lb of butter.

Cake Mixes

Cake	Cost
Betty Crocker	$0.97
Pillsbury	$0.97
Duncan Hines	$0.97
Generic econ.	$0.79

Amount of Supplies

Cake	Eggs	Water in cups	Oil in cups/butter
Betty Crocker	3 whites	1 and 1/4	1/3
Pillsbury	2	1 and 1/4	1/4
Duncan Hines	3	1 and 1/3	1/3
Generic econ.	3	2/3	1 stick butter

Cost of Supplies

Item	Cost	Amount
Butter	$0.29	1/4 lb = 1 stick
Eggs	$1.15	1 dozen
Icing	$1.15	16oz

Oil type	cost/48 oz	cost/gal
Wesson	$1.99	$5.59
Crisco	$2.19	$5.49
Generic	$1.89	$4.69

Task Two:

The company paying for the cake wants a bid for cleaning up the trash generated from making the monstrous cake. You want to make a profit of $1000 on this job.

Because you are environmentally conscious, you will make every effort to recycle as much of the waste as possible and provide the bakers with containers in which to sort the trash, i.e., separate receptacles for egg shells (NR), egg cartons (R) and cake boxes (R), plastic bags (NR) and oil bottles (NR), icing container (R), butter boxes and wrappers (NR): NR = Nonrecylcable and R= Recyclable. Cost at the landfill is $38/ton; however, trash dumping is prorated for each fraction of a ton, so it may be calculated proportionally. Unfortunately, you make no money on recycling.

Using the scales and materials provided, estimate the total amount of waste (in kilograms and pounds). Then find out the accumulated amounts for disposal and the accumulated amounts for recycling. Finally come up with your bid per pound (or per kilogram) to remove the trash. (Remember 1 pound = 2.197802 kg and 1 kg = .455 lb.) Show all work and justify your answer in a written proposal.

Extensions

- Find out how much cake and waste per person was created.
- Contact the local landfill to find out the charge for dumping nonrecyclable trash per pound or unit. Contact a recycling company to see if you could make any money from your recycling effort.
- Add more options for choosing ingredients for the cake, i.e., give prices for eggs either by size of egg or by different carton sizes.
- If you do not have access to a set of scales, the approximate weights are as follows:

Recycling Information

Item	Weight in grams
Empty Icing Container	35
Empty Oil 48	100
Cake box	41
Plastic cake mix bag	4.5
Egg shell	5.2
Empty Egg carton	47.2
Butter box and Wrapper	24

Remember:
1 lb. = 2.20 kg and 1 kg = .455 lb.

IDENTIFYING CONGRUENT PARTS

Subject: Geometry

Objectives: Given two congruent triangles, identify congruent parts. State and apply the SSS, ASA, SAS, AAS, HL, and HA Congruence Postulates. Construct triangles using these postulates as a basis. Use definitions and the congruence postulates to prove triangles congruent.

Materials: Students will need protractors, compasses, straight edges, construction paper, scissors, transparencies, and a collection of objects that are congruent such as a box of tooth picks, computer disks, rulers, quilt designs, plastic cups, golf balls, tessellations, Escher prints of tessellations, wallpaper with congruent designs, or parts to machinery.

Activities During the Lesson: (This lesson is designed using the cooperative learning technique "jigsaw." It may take more than one block depending on the length of the block.)

1. Students brainstorm and find examples of congruent objects in everyday life.

2. Groups are asked to become an expert on one of the triangle congruence postulates. The topic is assigned by the teacher.

3. Constructions illustrating the assigned postulate.

4. Find an example to illustrate the use of the postulate in a proof and present it to the class.

5. Members from each of these groups come together and form new groups, having an expert on each of the congruence postulates. They teach each other about their particular expert area.

6. Use the "Geometry Sketchpad" software for practice and for constructions.

Warm-Up Activity: The students are asked to find the number of triangles they see in a figure composed of different sizes of triangles and some of which overlap. This can be used to introduce the idea of congruent triangles and congruent figures making a natural transition into the lesson.

Lesson Outline:

- Teacher introduces the concept of congruence by showing examples of congruent figures (see materials). Students are then asked to brainstorm examples of congruent figures and to find examples of why they are important.

- Students are then placed into six groups and assigned a congruence postulate. Each group is to construct a triangle given the information in their congruence postulate. For example, the group assigned the

SSS Postulate would start with the lengths of three sides of a triangle and each group member would construct a triangle with the lengths of its sides the same as these. They are then to compare their triangles with one another. Are they different or are they the same? Would knowing that three sides of one triangle are congruent to three sides of another triangle be enough to say the triangles will always be congruent to each other? Are there any restrictions as to the lengths of the sides? Students are asked to make a conjecture based on their investigation and write it down.

- Each group is then given a statement to prove using this newly found information. They are to write out the proof on a transparency. (Each member should take good notes to use later in explaining the postulate to others.)
- New groups are formed by taking one member from each of the triangle congruence postulate groups and putting them together so that the new group has an expert on each triangle postulate.
- The new group members now take turns explaining their postulate using the construction, the conjecture, and the proof illustrating their individual postulate.
- The teacher should carefully monitor each group, both the original groups and the newly formed ones, to be sure that the information disseminated is correct and to give some direction if needed.
- To conclude the lesson one student is selected at random to present their understanding of one of the congruence postulates.
- The teacher helps to bring closure by using the transparencies with proofs on them and emphasizing the use of the postulates as well as by providing a comprehensive list of all the triangle congruence postulates.
- Students are taken to the computer lab during the next block and use the Geometry Sketchpad for reinforcement and practice. The teacher needs to prepare practice instructions and investigations for the students to use in the lab.

Lesson Extension or Activity: Students are to investigate the SSA situation and determine if it can also be used as a postulate to prove two triangles congruent. When can it be used and when can't it be used? They should use the construction techniques used in this lesson.

EXPANDING A RECYCLING PLANT

Title: The Power of Recycling

Subject areas: Algebra I, Algebra II, Math Analysis, Calculus

Brief description:

The student is asked to examine a proposal for expanding a recycling plant whose manager is predicting current and future linear growth. The student is expected to analyze given information and explore, using a graphing calculator, to determine whether the linear model proposed is the best model to predict future growth. The uniqueness of this set of data affords the student an opportunity to pose many different arguments. With careful work and justified reasons, the student could use a model in which growth can be represented by a quadratic, a cubic, an exponential, or a power function.

Grade levels: 9–12

Objectives:

Students will be able to:

- organize and analyze data
- use estimation skills
- justify and defend mathematical reasoning
- communicate mathematical ideas in writing

Background:

The students will need to know how to use a graphing calculator, not only for computation but also for graphing and statistical analysis. A knowledge of functions is essential so that the student knows which possible functions to explore. The students need to have experience justifying their responses.

Time required: It varies based on the course. This could be a culminating experience to do either in or out of the classroom.

Materials/resources: The student will need the task sheet and access to a graphing calculator.

Directions for task:

The primary purpose of this activity is for students to use technology (graphing calculators) to explore and compare different results using the same set of data. The student must then argue for his or her decision and support it with appropriate mathematical calculations. The report should contain at least one graph with an appropriate equation modeling the data. Also, the regression coefficient for his or her chosen data should be included to demonstrate that the student model is better than the plant manager's, or to defend the plant manager's choice. Language should be clear and to the point, and supported by evidence from the student's investigations.

The Power of Recycling

The manager of the city recycling plant contacted you to say that the amount of materials being processed has grown beyond the capacity of the plant. He has provided you with the records for the amount of materials processed during the past six years as shown in the table below. Also, he has given a graph of the data with his prediction and reasoning showing linear growth over the next five years and a correlation of .94 for his data. Included below are excerpts from his report.

Recycling at the City Plant

Year	Year Number	Millions of Pounds
1990	1	38
1991	2	42
1992	3	48
1993	4	57
1994	5	71
1995	6	99

Our city has increased its poundage of recycling from 38 million pounds to 99 million pounds in five years. After careful study, because our growth is clearly linear, I predict that our recycling efforts will surpass 134 million pounds by the year 2000. We need to plan for this now and expand our facilities.

Recycling at the City Plant **y = 11.5x + 19**

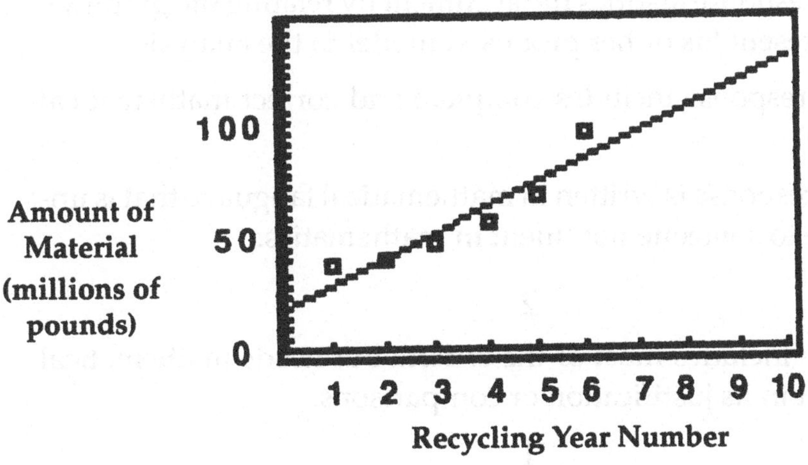

The regression coefficient is r = .94

As city manager, you are not sure that the model he has given you is the best one for the given data. In a report to the City Council, discuss the plant manager's prediction for future amounts of recyclable material and explain whether or not it is the best model.

In your report, be sure to give a detailed mathematical analysis supporting your results and request for expanded facilities. Include the graph you selected to represent your model. Also include the mathematical processes you used to make your estimates and predictions. Be sure to justify your graph in your analysis.

Scoring Rubric

4 = Task accomplished in exemplary manner
3 = Task accomplished successfully
2 = Task attempted with partial success
1 = Task attempted
0 = No response or off-task

Title: The Power of Recycling

4

- The student response includes all of the criteria of the task and exceeds the expectations by using more than one graph and equation to model the data and/or by including an exceptionally detailed analysis and justification in his or her report.

3

- The student response justifies the argument by relating the graph selected to represent his or her proposed model to the analysis.
- The student response includes complete and correct mathematical calculations.
- The student response is written in mathematical language that is understandable to someone not fluent in mathematics.

2

- The response includes most of the criteria but needs mathematical improvement in its justification or comparisons.

1

- The students response reflects little effort with little or no success and possibly a misunderstanding of the task.

0

- No response or off-task.

SIMULATING EXPONENTIAL DECAY

Title: Have Your Data and Eat It, Too!

Overview: In this lab, students are going to simulate exponential decay by using *M&M's Plain Chocolate Candies*. The student pairs will run trials removing M&M's from their population until no M&M's candies remain in the sample. Students will collect the data, enter it into graphing calculator lists, graph the data in a scatter plot on the calculator, determine a curve of best-fit from the scatter plot using the calculator utilities, and then use the information to answer questions. This activity works best when a large number of trials are performed and class data is complied at the end of the lab.

Objective: The students will determine the relationship between the "year" number and the "population" of candies. (This is a model of exponential decay with "year" as the independent variable and the number of remaining M&M's "m"-side up as the dependent variable.)

Level: Algebra II, Math Analysis/Precalculus, or Advanced Algebra (depends on the class and the teacher's objective for using the lab).

Materials:

- laboratory experiment data collection sheets
- graph paper (if teacher wants data plotted by hand)
- M&M's Plain Chocolate Candies
- navy beans to replace removed M&M's
- small plastic cup for holding candies or a small plastic bag for each group
- paper plate for each group for "dumping" candies
- graphing calculator

Teacher Instructions:

Put students into groups of two or three. (You need lots of data sets so that the experiment will model exponential decay better.) Give each group a cup or a plastic bag with 50 to 70 M&M's, a cup with 50 to 70 navy beans, and a data collection sheet.

Have each group count and record the number of candies in their cup/bag and list it as Year 0. Then have them put the candies in the cup/bag and shake it gently to mix the candies. Then they should pour the candies onto the paper plate, and remove the M&M's that have the m-side down and count and record the number left (m-side up) as the population for Year 1. They should replace the M&M's removed with navy beans. (Half-life means that the molecules have changed ion-level, not disintegrated, so the decayed M&M's should be replaced so that students understand the concept. Matter is neither created nor destroyed.) The students repeat this process until no M&M's are left and only navy beans are left.

The teacher needs to record all class data so that students can transfer it to the master table on their sheets for graphing.

The students then use the graphing calculator to determine an exponential equation of best fit and answer questions using the equation.

Have Your Data and Eat It, Too!

Group Members' Names

____

____

____

In this experiment, you are going to perform a mathematical investigation using *M&M's Plain Chocolate Candies*. You need to pick up a cup/bag of candies and a paper plate. Also, you will need to record your data for both your group and the entire class in the tables on this sheet. Carefully follow the directions below.

Procedure:

(1) Count the number of M&M's in your population and record it for Year 0. This is the initial population.
(2) Put the candies back into the cup/bag and shake gently to mix them.
(3) Pour them onto the paper plate.
(4) Remove the M&M's which have the m-side down.
(5) Replace the M&M's you removed with the same number of navy beans.
(6) Count and record the number of M&M's left on your plate.
(7) Repeat steps 2 through 6 until no M&M's remain.

Your Individual Data

Year	# in Population
0	
1	
2	
3	
4	
5	
6	
7	
8	
9	
10	
11	

Cumulative Class Data

Year	# in Population
0	
1	
2	
3	
4	
4	
6	
7	
8	
9	
10	
11	

Graph your data points for the class data on graph paper. *Exclude* the last year when there are no M&M's left.

Answer these questions:

1. Describe what the graph of your data looks like. What type of equation do you think would best describe this set of data?

2. If this experiment were a model of radioactive decay, what would the half-life of that element or isotope be? (According to your data, about how many "dumps" would it take for you to end up with half of your initial population?)

   Why would we exclude the last year when the population is zero?

3. (a) Using a graphing calculator, enter the Year Number into List 1 and the Population Number into List 2. (Exclude the last year when the population reached 0.)
   (b) Set an appropriate window for your graphing calculator and then graph your data in a scatter plot.
   (c) Use Exponential Regression to determine the equation for your data.
   - Write the equation of your curve here. ____________________
   - Transfer this information to your equation graphing screen and have the calculator draw the graph onto your scatter plot.
4. Choose two or three of your data points and apply the growth/decay formula $\mathbf{y = Ca^x}$ to write an equation for your data points.
5. Describe how well your data fits the equation you found in Question 3. Graph the equation you found in Question 4 and describe how well your data points fit it.

QUADRATIC RELATIONSHIPS

Title: Pendulum Lab

Objective: Students determine the relationship between the length of a pendulum and its period. (This application is a quadratic relationship.)

Level: Algebra II, Trigonometry (if extensions are used)

Materials:

- washers or nuts to which a string may be attached (1 per group)
- kite string or yarn (enough for each pair of students to have 4 different lengths; approximately 2 to 3 m per group)
- watch or clock with a second hand (1 per group)
- lab sheet to record data (1 per student)
- meter sticks or tape measures (1 per group) and protractors (1 per group) (metric measurements are used to parallel science measurements)
- writing utensil
- graphing calculator
- newsprint marked along the top with appropriate time intervals on which to tape student's strings from #18 in the student activity

Teacher Instructions:

- The accompanying lab contains directions so that the students may conduct the lab themselves if you provide them with the materials; it will free you to be the lab assistant walking around answering questions.
- This lab works best in pairs so that each person is doing and not watching. Have them switch parts: while one swings and counts the other keeps track of time and records data. This will also work fairly well with groups of three in which one person is the swinger and counter, the second person is a verifying counter, and the third person is the timer.
- The students can graph the data by hand or by using the graphing calculator. Also, you can choose to let the calculator determine the curve of best fit or create it by hand using a difference table and systems of equations.

- Remind the students to be careful and as accurate as possible with their measurements and timing to reduce laboratory error. Results will probably not be perfect but a perfect data set can be given to the students later so that they can see where their errors occurred.
- **WARNING:** If you are a teacher who cannot deal with classroom noise, then this activity is probably not for you, because the students must be up and about doing things. You could have them conduct the investigation by giving them a "canned" set of data and have them work individually. However, this is not nearly as powerful an experience!

PENDULUM LABORATORY

	Length	Time Elapsed	Number of Cyles	Period (T)	T/L	T^2	T^2/L	T^3	T^3/L
1									
		Average							
2									
		Average							
3									
		Average							
4									
		Average							

Name: ______________________

Partner: ______________________

"Hickory, Dickory, Dock, the Mouse Ran Up the Clock..."

Objective: To determine the relationship between the length of a pendulum and its period.

Data Collection Instructions

1. Fasten your weight to your string. Cut off excess string close to your knot.
2. Mark **four** different lengths along your string with at least 20 to 30 cm between each mark.
3. For the first mark along your string, use your protractor to draw the string along the protractor to about 30 degrees. Hold the protractor away from the string to eliminate friction and drag. Holding your string taut, let the string go and count the number of back and forth swings (i.e., cycles) in 20 seconds. Record the data in your table. **Note: 1 cycle = 1 back and forth swing**
4. Repeat this process **four** more times, recording your data each time in the chart.
5. Repeat steps three and four for your other three marks on the string and record your data in the table.
6. Be sure you have recorded the length of your string for each group of swings and 20 seconds for your time elapsed.

Data Analysis Instructions

7. Find the average of the five cycles for each string length.
8. Find the period (**T**) by dividing the time elapsed by the average number of cycles. Record this data.
9. Find the ratio: **T** (period) ÷ **L** (length) for each set of data. Record it.
10. Find T^2 / L for each set and record it.
11. Find T^3 for each set and record it.
12. Find T^3 / L for each set and record it.
13. Draw graphs for #9 through #12, letting the **x-axis be the length of your string** and the **y-axis be the ratio you found for each calculation**.
14. Using your graphs and your lab sheets see if there are any trends or relationships within your data that extend to all four different string lengths.
15. If not, do you see any column which might contain a relationship were it not for experimental error?

16. In the data table at right, fill in your string lengths and then calculate the period **T** of a pendulum using the formula $T \approx 2\pi\sqrt{\frac{L}{g}}$, where **L** is the length of your string and **T** is the period of your pendulum according to the formula and **g** is the effect of gravity, **9.8 meters/second²**.

Length	Period

17. Using this data, recalculate T^2/L in the table at the right. What do you notice about this data now?

    What is the value of $\frac{4\pi^2}{g}$?

    Show that $\frac{T^2}{L} = \frac{4\pi^2}{g}$.

T^2/L

18. Cut string lengths to match the four lengths you chose for your experiment from the string provided. Tape the strings to the newsprint provided **above** the length of the period you got for each string. Do you notice the shape of a familiar curve with all strings hung?

    Are there any outliers in the data?

19. Complete the table of values with **T**, the period of the pendulum as the independent variable and $f(T) = 4.11T^2$ as the dependent variable. From what is the coefficient of T^2 derived?

T	0.5	1	1.5	2	2.5	3	3.5	4
f(T)								

(An option is to give the students the table and have them derive the equation by using difference tables to create equations which can be solved using linear systems of equations.)

A Pendulum Problem Set

The time it takes for a pendulum to swing back and forth and return to its starting position is called the period (T) of the pendulum. The period can be calculated using a radical equation, where **L** is the length of the pendulum and **g** is the acceleration due to gravity:

$$T = 2\pi \frac{L}{g}$$

Calculate the answer to the following problems to the nearest tenth. Use π = 3.14 unless you have a calculator with a π key.

1. The pendulum of a grandfather clock has a period of 2 seconds. How long is the pendulum in meters? (Assume that on earth, acceleration due to the force of gravity is 9.6 meters/second^2.)
2. The pendulum of an antique wall clock makes one complete swing every 1.5 seconds. What is the length of its pendulum?
3. The pendulum of a cuckoo clock is 6 cm long. How many full swings does the pendulum make in 2 seconds? (Remember to change meters to centimeters.)
4. Acceleration due to gravity is only 1.6 m/s^2 on the moon. If the grandfather clock from Problem 1 is taken to the moon, how will the period be affected?
5. Suppose the same grandfather's clock is taken to a planet where acceleration due to gravity is four times that of the value on earth. What will happen to the period of the pendulum?
6. Suppose the cuckoo clock in Problem 3 is taken to the moon. How many full swings would the pendulum make in 6 seconds?
7. The pendulum of a small antique clock makes one complete swing every 3 seconds. It moves half as fast as the clock in Problem 2. How does the length of the pendulum compare to the pendulum in the clock in Problem 2?

These problems were adapted from Prentice Hall's Algebra 2 with Trigonometry Teacher Resource Materials *by Hall and Fabricant.*

EXTENSION

Read Edgar Allan Poe's "The Pit and the Pendulum" and discuss the pendulum in the story. Write some problems that the class could solve using the pendulum in the story.

OPTIMIZATION PROBLEMS

Subject: Calculus or Precalculus

Objectives: Students will be able to set up and solve an optimization problem.

Materials: A wrapper from a Hungry Jack biscuit can, construction paper, rice or beans, tape, scissors, cardboard or stiff paper like folders, tape measures, rulers.

Prior Assignment: Students have done related rates problems and know how to find the derivative of a function with respect to a stated variable. Students have been asked to bring in a couple of cans of food from home. They may take them back later or they may donate them to a canned food drive.*

Activities During the Lesson: (This lesson is designed for one block of time.)

1. Groups of two will make right circular cylinders out of parallelograms copied from the wrapper from a Hungry Jack biscuit can.

2. The students will then make a guess as to which of the two cans formed will hold the most rice. Which one has the greatest volume? They will then use the measuring instruments and determine the volume of each can and its total area.

3. Using the cans brought from home the students will brainstorm what considerations they think go into designing a can to contain a food item.

Lesson Outline:

(This lesson was designed as the introduction to optimization maxima/minima problems and setting the stage for applications related to these types of problems.)

- Working in pairs, students are given two Hungry Jack biscuit wrappers or traced copies. Students are instructed to roll the wrappers (parallelograms) back up to make two different sizes of cylinders. Tape these two cylinders to hold their shape. One is shorter and fatter. The other is longer and thinner.

- The teacher should have two cylinders of the two different sizes prepared as discussion models to expedite the student activity. Problem: determine if the two cylinders hold the same amount. Students find the volume of each cylinder by measuring the necessary dimensions of the cylinders (or use rice or beans to fill each and compare the amounts). Once the volumes have been determined asked the students how this might impact the producer of a product sold in these two types of cans having the same lateral surface area. Discuss what dimensions might be more profitable for the manufacturer to use.

- Introduce the idea of maximizing and minimizing a quantity. Teacher-directed development of the procedure to analyze, set up, and work an optimization problem. Use graphs of functions students have graphed using information from the first derivative (relative extrema).
- Using the cans of food that the students have brought to class make observations concerning the sizes of the cans and the amounts contained in the cans. Discuss what factors might be used to decide on the amount contained in each can. How is it decided? Why are the cans of varying sizes?
- After presenting the students with the following problem use the procedure above and work the problem together or in their groups. Problem: A soup company wants to make soup cans to hold a given amount with the least amount of manufacturing material. How would soup cans be designed to maximize profit? What factors must be considered? Using this problem the teacher reinforces problem-solving strategies for mathematically determining maximum and minimum values under given conditions. Upon arriving at the conclusion that the ideal can, one that would hold a given volume and have the minimum surface area, would have the diameter of the can equal to the height of the can, ask students to look at all of the cans brought to class and see if any of them fit this description. Surprisingly to the students there are almost no cans that have this relationship between the diameter and the height of the can. Why? To address this question refer to the lesson extension.
- Working in their groups the students will design a maxima/minima problem that exists in their own experience. For example, the SGA wants to sell tickets to the faculty talent show to groups to get the most people to attend, and also to make the most money. If they given a discount of 20 cents for each number of tickets sold over 10 tickets with a single ticket price of $5.00, how many must be sold to maximize profit? The students write up the problems and their solutions. Practice problems should also be assigned.
- At the next class meeting the students exchange problems, work them , and check their answers. More problems may be assigned as needed for practice.

Lesson Extension or Activity:

As a culminating assignment the students write a letter to the manufacturer of the food item that they brought in for class. The letter is written to inquire about the reason why the ideal can dimensions are (or are not) used by the company in the production of cans for its food items. All letters are proofed by the teacher for clarity , completeness, and correctness. These are mailed to the companies and shared as responses arrive. All letters are posted on the bulletin board after they are read to the class.

* Students have been instructed to bring a canned food item to class for their home enjoyment assignment. Traditionally in the AB Calculus class, we study optimization problems shortly before Thanksgiving. Our school has a canned food drive for Thanksgiving food baskets to be given to a local service organization for distribution to families in need of assistance. We donate our canned food items to this food drive after we have done our class investigation.

APPENDIX B

RESOURCES

NCTM MATERIALS

National Council of Teachers of Mathematics, 1906 Association Drive, Reston, VA 20919-1593; Phone: (703) 620-9840; Fax: (703) 476-2970; Orders only: (800) 235-7566; Web site: http://www.nctm.org

Curriculum and Evaluation Standards for School Mathematics

Professional Standards for Teaching Mathematics

Assessment Standards for School Mathematics

The complete set of volumes which set the vision for reshaping mathematics education for today and the future. Examples are given which support the discussions.

Activities for Engaging Learning and Teaching Selections from the Mathematics Teacher, edited by Christian R. Hirsch and Robert A. Laing

Activities for middle school and high school arranged by mathematical discipline. Students use manipulatives, calculators (scientific and graphing) and computers. Complete directions are given for teachers and students with reproducible activity sheets.

Activities for Junior High School and Middle School Mathematics: Readings from the Arithmetic Teacher and the Mathematics Teacher, edited by Kenneth Easterday, Loren Henry, and F. Morgan Simpson

Activities organized by strands including counting and place value; decimals; fractions; percents; probability and statistics; and problem solving.

Activities from the Mathematics Teacher, edited by Evan M. Maletsky and Christian R. Hirsch

Discovery lessons, laboratory experiences, games and puzzles, and model constructions for grades 7 through 12 with reproducible pages.

Algebra for Everyone, edited by Edgar l. Edwards, Jr. and sponsored by the Mathematics Education Trust

Set of essays discussing specific aspects of teaching algebra to all students and addressing issues such as "problems facing underachieving populations."

Algebra for Everyone In-Service Handbook, David J. Glatzer and Stuart A. Choate and sponsored by the Mathematics Education Trust

Activities for educators who will teach the students who have been historically low achievers and/or a part of the underserved populations.

Algebra in a Technological World: Addenda Series, Grade 9-12, M. Kathleen Heid, Jonathan Choate, Charlene Sheets, and Rose Mary Zhiek

Addresses the teaching and learning of high school algebra following the vision of the NCTM *Standards* and the changes brought about by graphing calculators and computer software.

Applications for Secondary School Mathematics, Readings from the Mathematics Teacher, edited by Joe Dan Austin

Collection of articles that emphasize the use of real-world applications to motivate students to apply mathematical history. Essays are grouped by mathematical discipline.

Connecting Mathematics: Addenda Series, Grades 9–12, edited by Gary Froelich with Kevin G. Bartkovich and Paul A. Foerster

Links content proposed by the *Standards* to current algebra and geometry programs. Classroom-ready examples illustrate how new content can be integrated into the curriculum connecting traditional topics that may seem isolated otherwise.

Connecting Mathematics Across the Curriculum (1995 Yearbook), edited by Peggy A. House

Illustrates how mathematics can be connected to other mathematics, to other subjects of the curriculum, and to the everyday world.

A Core Curriculum—Making Mathematics Count for Everyone: Addenda Series, Grades 9–12, by Steven P. Meiring, Rheta N. Rubenstein, James E. Schultz, Jan de Lange, and Donald L. Chambers

Offers several different curriculum models for organizing mathematics content as envisioned by the *Standards.* Includes sample syllabi, lessons, and activities that help frame the curriculum to develop mathematical power and technical competence.

Data Analysis and Statistics Across the Curriculum: Addenda Series, Grades 9–12, Gail Burrill, John C. Burrill, Pamela Coffield, Gretchen Davis, Jan de Lange, Diann Resnick and Murray Siegel

Examples and activities that illustrate how to integrate statistical concepts into the standard high school curriculum.

Dealing with Data and Chance: Addenda Series, Grades 5–8, Judith S. Zawojewski with Gary Brooks, Lynn Dinkelkamp, Eunice D. Goldberg, Howard Goldberg, Arthur Hyde, Tess Jackson, Marsha Landau, Hope Martin, Jeri Nowakowski, Sandy Paull, Albert P. Shulte, Philip Wagreich and Barbara Wilmot

Builds upon the people's natural understanding and daily use of data and chance. Activities illustrate data gathering, communication, problem solving, reasoning, and connections.

Developing Number Sense in the Middle Grades: Addenda Series, Grades, 5-8, Barbara J. Reys with Rita Barger, Maxim Bruckheimer, Barbara Dougherty, Jack Hope, Linda Lembke, Zvia Markovits, Andy Parnas, Sue Reehm, Ruthi Sturdevant, and Marianne Weber

Ideas and activities for building lessons to develop number sense in the curriculum.

Discrete Mathematics across the Curriculum, K–12 (1991 Yearbook) edited by Margaret J. Kenney

Develops a vision of discrete mathematics in the K-12 curriculum and extends that vision to college mathematics. Articles explore graph theory, matrices, counting methods, recursion, iteration and induction, and algorithms. Activities and teaching units for secondary teachers are included.

Exploratory Problems in Mathematics, by Frederick W. Stevenson

This book provides a compendium of open-ended problems which encourage students to be creative in their mathematics thinking and exploration. Problems can be worked on individually or in groups. Ideas are included for getting started but no one answer is given—teachers need to be creative, too!

Fractals for the Classroom

Part One: Introduction to Fractals and Chaos

Part Two: Complex Systems and Mandelbrot Set by Heinz-Otto Peitgen, Hartmut Jurgens, and Dietmar Saupe

Two volumes which introduce the dynamic concepts of chaos theory and fractal geometry and their relationships to each other, to other aspects of mathematics and to natural phenomena.

Fractals for the Classroom: Strategic Activities, Volumes 1 & 2 by Hartmut Jurgens, Evan Maletsky, Heinz-Otto Peitgen, Terry Perciante, Dietmar Saupe, and Lee Yunker

Hands-on activities which enable students to discover the underlying mathematical principles and characteristics of fractals. Students are amazed at their discoveries and turn on to a branch of mathematics that is rarely covered in the traditional classroom.

Geometry from Multiple Perspectives: Addenda Series, Grades 9–12 by Arthur F. Coxford, Jr. with Linda Burks, Claudia Giamati, and Joyce Jonik

Links the geometry content proposed in the *Standards* to current geometry programs. Includes classroom-ready activity sheets with instructional suggestions.

Geometry in the Middle Grades: Addenda Series, Grades 5–8 by Dorothy Geddes with Julianna Bove, Irene Fortunato, David J. Fuys, Jessica Morgenstern, and Rosamond Welchman-Tischler

Focuses on two- and three-dimensional geometric concepts, relationships among properties of shapes, transformations, and geometry enrichment activities. This book includes both assessment and technology ideas.

Historical Topics for the Mathematics Classroom

This NCTM yearbook offers extensive historical information grouped by topics with articles included on almost every topic that might interest you or your students. The overviews and detailed capsules set the corpus of mathematics into a proper framework so that an appreciation can be gained for the development of mathematics from its earliest beginnings.

How to Evaluate Progress in Problem Solving by Randall Charles, Frank Lester, and Phares O'Daffer

Includes ideas that help teachers integrate problem solving into the curriculum and helps the teachers and students determine what is being evaluated and how to go about evaluating it. The authors describe how to manage an evaluation program and use the results.

How to Use Cooperative Learning in the Mathematics Class by Alice F. Artzt and Claire M. Newman

Cooperative learning activities are included for elementary through secondary students with ideas about how to implement cooperative learning into the classroom and its benefits for all students.

The Ideas of Algebra, K–12 (1988 Yearbook) edited by Arthur F. Coxford

Comprehensive volume describing the accessibility of algebra to all students with emphasis on equations and expressions, word problems, and the use of technology in algebra classes. Includes teacher-tested ideas for teaching algebra.

Ideas: NCTM Standards-Based Instruction, Grades 5–8 complied and edited by Michael C. Hynes

Forty-seven lessons compiled from the "Ideas" department of the 1991–94 *Arithmetic Teacher.* Lessons include reasoning, connections, communication, and problem solving.

Imaginative Ideas for the Teacher of Mathematics, Grades K-12: Ranucci's Reservoir edited by Margaret A. Farrell

Enrichment activities cover five areas: patterns, mathematics in the world around us, spatial visualization, inventiveness of geometry, and games to learn by.

Learning and Teaching Geometry, K-12 (1987 Yearbook) edited by Mary Montgomery Lindquist

Twenty articles about the way students learn geometry, applications, activities, incorporating geometry into other areas of mathematics, and preparing teachers to teach geometry.

Mathematical Modeling in the Secondary School Curriculum: A Resource Guide of Classroom Exercises edited by Frank Swetz and J.S. Hartzler

Discusses how to use mathematical modeling in the mathematics classroom. Includes modeling exercises from general mathematics through calculus with interdisciplinary ideas.

Mathematics Assessment: Myths, Models, Good Questions, and Practical Suggestions edited by Jean Kerr Stenmark

Step-by-step examples of alternate assessment ideas including portfolio assessment. Ideas developed by teachers for teachers.

Mathematics Teacher (MT)

Monthly publication by NCTM that is a must for grades 8–12 classroom teachers. Ready-to-implement lessons in each volume, a monthly calendar, which is an excellent source of warm-up problems, "technology tips," "sharing teaching ideas," "media clips," and reviews of new products.

Mathematics Teaching in the Middle School (MTMS)

The middle school version of the *Mathematics Teacher* valuable for teachers in 6-12. Each issue has appetizing menus of problems for students...similar to the calendar in MT. Contains reviews of new materials and software and three sections for students.

Measurement in the Middle Grades: Addenda Series, Grades 5–8 by Dorothy Geddes and others

Full of sample activities that relate to the real world. Includes easily reproducible activities that focus on estimation and high-order thinking skills.

Patterns and Functions: Addenda Series, Grades 5–8 by Elizabeth Phillips with Theodore Gardella, Constance Kelly, and Jacqueline Stewart

Activities to develop understandings of concepts of exponents, number theory, rational numbers, measurement, geometry, probability, and functions.

Projects to Enrich School Mathematics, Levels 1,2 and 3

Independent study projects to challenge and enrich average and above-average mathematics students.

Teaching and Learning Mathematics in the 1990s (1990 Yearbook) edited by Thomas J. Cooney

Explores the changing roles of teachers and students in a decade of the *Standards*. Suggests effective methods for teaching and assessment.

Teaching with Student Math Notes (Volumes 1 and 2) edited by Evan Maletsky

Reproductions of *Student Math Notes* published every other month in the *NCTM News Bulletin*. Ready to be reproduced, the topics covered are wide and varied. Answers are included with each activity.

AIMS MATERIALS

AIMS ("Activities Integrating Mathematics and Science") Education Foundation, P.O. Box 8120, Fresno, CA, 93747. Materials to enrich the education of students K-12 through hands-on activities that integrate mathematics, science, and other disciplines. High school teachers should not ignore these resources because many are directly applicable to either the algebra or geometry curriculum while others are easily adaptable. A problem of the month is available on its Web site. Each volume is $14.95 plus 10% shipping and handling.

Soap Films and Bubbles (5–9)

Exploration into the behavior of soap films (check for Rocky's stuff).

Math + Science, A Solution (5–9)

Use everyday materials such as marbles, M & M's candies, old shoes, and rubber balls to integrate mathematics and science.

Down to Earth (5–9)

Discovery activities in geology, meteorology, and oceanography. One lab lets students discover the differences between the Colorado River and the Mississippi River: the Colorado River dug the Grand Canyon while the Mississippi River keeps on rolling.

Our Wonderful World

Environmental explorations to be conducted out-of-doors.

Out of the World

Astronomy and solar system, activities.

Pieces & Patterns: A Patchwork in Math & Science

Activities covering a variety of topics. Mirrors and kaleidoscopes let students develop concepts from geometry.

Floaters and Sinkers

Hands-on activities dealing with mass, volume, and density.

Through the Eyes of the Explorers

Map-making activities make the Lewis and Clark expeditions come alive. Mathematical concepts progressively become more complex. Estimation, angle measurements, and linear measurements are explored extensively.

Historical Connections in Mathematics, Volumes I, II, and III

These volumes include biographical information, famous quotations, problem-solving exercises, and hands-on learning experiences. The lives of 30 mathematicians are profiled, 10 in each volume.

What's Next? A Pattern Discovery Approach to Problem Solving
Volumes 1,2, and 3

Pattern recognition problems which are easily adaptable to a wide-range of abilities and interests. Challenges students to develop abstract formulas from concrete data.

DALE SEYMOUR MATERIALS

Dale Seymour Publications, P.O. Box 10888, Palo Alto, CA, 94303. Toll Free Number: (800) 872-1100

Finite Differences by Dale Seymour and Margaret Shedd

Generalizing problems arising from sequences and tables with ideas for determining methods for finding equations to represent the data through difference tables, blackline masters.

THE MESA SERIES, University of Washington Mathematics, Engineering, Science Achievement Group. Middle School Replacement Units which can be modified to use in secondary courses. Activity oriented, with many open-ended questions.

Classifying Fingerprints

Measuring Dinosaurs

Packaging and the environment

In the Pharmacy

Investigating Apples

Measuring Earthquakes

In the Air

Algebra Experiments I & II: Exploring Linear Functions and Exploring Nonlinear Functions by Mary Jean Winter and Ronald J. Carlson

Two laboratory books with algebra experiments complete with lab activity handouts and teacher guidelines for implementation. Equipment and material needed are easily obtained: cereal, miniature model cars, a stopwatch, cups and lids, etc. May be put directly into the curriculum.

Geometry Experiments: Exploring Algebraic Connections by Mary Jean Winter and Ron Carlson

Geometry laboratory experiments which incorporate geometric concepts with algebraic representation and integration of graphing calculators. These are complete with lab activity handouts and teacher guidelines for implementation. May be directly put into the curriculum.

The Language of Functions and Graphs by Shel! Centre for Mathematical Education

This British publication describes functions and their applications in depth, for incorporating the function concept throughout the curriculum, and includes a set of blackline masters.

Algebra in the Real World by LeRoy Dalton

A collection of applications that illustrates examples of mathematics found in the real world. Lessons for second year algebra incorporate traditional concepts in applicational situations.

Qualitative Literacy Series: Tested by the American Statistical Association and the NCTM Joint Committee on the Curriculum in Statistics, this series provides current topics and examples for teaching probability and statistics to secondary students. Student editions are sold separately from the teacher edition and may not be reproduced.

Exploring Data by C.M. Newman, T.E. Obremski, and R.L. Schaeffer

Exploring Probability by M. Gnanadesikan, R.L.Scheaffer, and J. Swift

The Art and Techniques of Simulation, J.M. Landwehr, J.Swift, and A.E. Watkins

Exploring Measurement by J.M. Landwehr and A.E. Watkins

Exploring Surveys and Information from Samples by P.Barbella, J. Kepner, and R. Schaeffer

There is also a videotape available.

Mathematical Investigations: Book One, Two and Three by R. Souviney, M. Britt, S. Garguilo, and P. Hughes

Activities that incorporate real world applications and make connections to other disciplines. Activities include some for algebra, geometry, graphing, counting, probability, and statistics in problem solving. May be integrated into whole class or individual activities.

Algebra Warm-Ups by Scott McFadden

Seventy sets of warm-up lessons which can be completed in about 10 minutes to introduce new concepts or review previous material. Blackline masters.

Algebra Problems, One Step Beyond by Reuben Schadler

Thirty-three sets of reproducible problems to include in a first-year algebra course. Problems include warm-ups, lesson, extension, and follow-ups.

Flatland by Edwin A. Abbott

This is a delightful book about living in a two-dimensional world. It is wonderful companion reading for any geometry course and lends itself to many projects.

Graphing Calculator Activities, Exploring Topics in Algebra I and II by Charles Lund and Ewin Andersen

Reproducible masters for investigation the graphs and functions common to Algebra I and Algebra II. References for most popular graphing calculators.

Get It Together by Tim Ericson

Problem solving presented in a cooperative learning environment. Groups of students get clues to the solution for a problem. Each student is dependent upon the other students' information and they must work together to determine the solution.

CREATIVE PUBLICATIONS MATERIALS

Creative Publications, Order Department, 5623 W. 115th Street, Worth, IL 60482–9931; Toll Free: (800) 624-0822

Middle School Math with Pizzazz (Grades 5–8) by Steve and Janis Marcy

Five binders with puzzles keyed to the mathematics objectives from grades 5 through 8. Operations on whole numbers, basic facts, numeration, decimals, operations, applications, problem solving, fractions, percents, perimeter, angles, area, square-roots, and probability and statistics. Each binder may be bought separately.

The Problem Solver with Calculators for Grades 4–8 by Goodnow, Hoogeboom, Moretti, Stephens, and Scanlin

Problem solving strategies in a three-section format: teaching strategies, practice problems on blackline masters, and solutions sections.

Pre-Algebra with Pizzazz (Grades 7–10) by Steve and Janet Marcy

Reproducible activity pages which are set up in a puzzle format for pre-Algebra objectives. Activities are sequenced and divided into sections.

Algebra with Pizzazz (Grades 8–12) by Steve and Janet Marcy

Riddles and joke-motivated reproducible pages sequenced and organized into sections which include some pre-algebra objectives. Even Geometry and Algebra II students enjoy returning to this format.

KEY CURRICULUM MATERIALS

Key Curriculum Press, P.O. Box 2304, Berkeley, CA 94702–0304; Toll Free: (800) 995-MATH (6284); Toll Free Fax: (800) 541-2442; Web site: http://www.keypress.com

Discovering Geometry: An Inductive Approach by Michael Serra

Innovative textbook actively involves students in their own learning. Investigations, constructions, activities, and projects are included. Students experiment to make their own geometric conjectures. Extensive resources for the teacher.

Advanced Algebra Through Data Exploration: A Graphing Calculator Approach by Jerald Murdock, Ellen Kamischke, and Eric Kamischke

Innovative textbook that combines data exploration, hands-on activities, and graphing calculators to enhance advanced algebra content. Projects and methods for alternative assessment are provided. Extensive resources for the teacher.

Interactive Mathematics Program (IMP) by Dan Frendel, Diane Resek, Lynne Alper, and Sherry Fraser

Four-year program to "reinvigorate high school mathematics." The curriculum integrates science, history, and literature into the mathematics curriculum. One unit from the year-one program, "The Pit and the Pendulum," uses the Edgar Allan Poe tale to set a study of concepts that include normal distribution, measurement variation, standard deviation, curve-fitting, and families of functions. Extensive resources for the teacher.

Calculus: Concepts and Applications by Paul Foerster

This textbook was specifically written for high school AP calculus teachers by a high school teacher. Written with the assumption that every student has a graphing calculator available at all times, the material includes cooperative learning, projects, and journal writing. Extensive resources for the teacher.

Problem Solving Strategies: Crossing the River with Dogs by Ted Herr and Ken Johnson

Excellent textbook to teach problem solving in the classroom. The approaches are multidimensional and students gain confidence in choosing which strategy works best. Allows the student to see that there are many ways to solve one problem.

The Geometer's Sketchpad 3

This is geometry software for Euclidean, coordinate, transformational, analytic, and fractal geometry.

Add-on packages include:

Exploring Geometry with the Geometer's Sketchpad

Geometry through the Circle with the Geometer's Sketchpad

Pythagoras Plugged In: Proofs and Problems for the Geometer's Sketchpad

Exploring Conic Sections with the Geometer's Sketchpad

Perspective Drawing with the Geometer's Sketchpad

Exploring Trigonometry with the Geometer's Sketchpad

TesselMania

This software's simple interface allows students to investigate Escher-like tessellations. It comes with a book of blackline masters specifically designed for secondary instruction.

HRM VIDEO MATERIALS

HRM Video, 175 Tompkins Avenue, Pleasantville, NY 10570; Toll Free (800) 431-2050; Fax: (914) 747-1744; Web site: www.hrmvideo.com

Applications of Logarithms (Grades 9–12)

A kit that includes a video, blackline masters, and overhead transparencies. Emphasis is on real-world applications.

Applications of Trigonometry (Grades 9–12)

A kit that includes a video, blackline masters, and overhead transparencies. Emphasis is on real-world applications.

Applications of Conic Sections (Grades 9–12)

A kit that includes a video, blackline masters, and overhead transparencies. Emphasis is on real-world applications.

SUNBURST MATERIALS

Green Globs and Graphing Equations designed by Sharon Dugdale and David Kibbey

There are four environments in the game: *Equation Plotter, Linear and Quadratic Graphs, Green Globs,* and *Tracker.* In *Green Globs,* 13 green dots are randomly "splattered" on a coordinate plane and students work to define the function that will best hit as many globs as possible in a game environment.

The Geometric PreSupposer, The Geometric Supposer Series, and *The Geometric Super-Supposer* designed by Education Development Center, Dr. Judah L. Schwartz, MIT and Harvard, and Dr. Michal Yerushalmy

Provides students an opportunity to discover concepts, relationships, and theorems of Euclidean geometry, and to extend their findings to make their own discoveries.

OTHER MATERIALS

The Joy of Mathematics

More Joy of Mathematics

The Magic of Mathematics all by Theoni Pappas

Wide World Publishing/Tetra, P.O. Box 476, San Carlos, CA 94070

These three books contains numerous examples, descriptions, and insights into the wonders of mathematics. They give teachers great creative ideas for making connections in the mathematics classroom.

The Great Book of Math Teasers by Robert Muller

Lateral Thinking Puzzlers by Paul Sloane; illustrated by Myron Miller

Baker Street Puzzles by Tom Bullimore

Sterling Publishing Co., Inc., 387 Park Avenue South, New York, NY 10016

These three books provide a variety of nonroutine problems which can be used for warm-ups, extensions, or end of class activities to promote problem-solving and creative thinking.

The Mathematical Recreations of Lewis Carroll: Pillow Problems and *A Tangled Tale* by Lewis Carroll

Dover Publications, Inc., New York, NY

Numerous puzzles posed by Lewis Carroll. These are good for longer periods of time than warm-ups. They range from those which can be solved using arithmetic to those solvable using algebra, geometry, trigonometry, and calculus.

The Buck Book: All Sorts of Things to Do with a Dollar Bill—Besides Spend It by Anne Akers Johnson

Klutz Press, 2121 Staunton Court, Palo Alto, CA 94306 (415) 857-0888

This book provides activities for folding a dollar bill into various "origami" figures. It is a great way to teach fractions, percent, and geometry by posing appropriate questions as you teach the students how to do the folds.

This list of resources is by no means complete. We are providing you with a starting place for resources that we use in our classrooms to build our units and lessons.

APPENDIX C

WEB SITES

If you haven't looked on the Internet for resources to use in your math class, you have really missed something. The Internet is full of lesson plans, ideas, information, math history, puzzles, contests, games, and homework help for all the math topics and subjects taught in the middle school and high school math classrooms.

Most teachers are still in the throws of planning new lessons to include the *Standards*. We are continuously trying to update our curriculums to include technology, applications, problem solving, and real-world math, and trying to do it creatively while still emphasizing skills. This often leaves little time to browse on the Internet and visit sites to find those that have something we can use in the classroom.

This section provides you with sites you can visit and find material to integrate into your lesson plans. We have visited each of these sites and found them to be worthwhile. We have not put them in any special order and hope that all sites are still up and running. In addition to a Web site's address, we have annotated it so that you will know what the site is about. Thus, you can select the sites that best meet your needs. Hopefully, this list will save you time when trying to locate materials or information on a specific math topic. This list is by no means complete, but it is a start. You will also find an annotated list of sites done by our students as part of a project for our classes at the end of our list.

Classroom Connect
http://www.classroom.net

> One of the very best sources we found for math educational Web sites is a journal called *Classroom Connect*. We asked our media center specialist to order it for our library so that all of our teachers could have access to it. It is for teachers of all subjects and grades. We recommend it for all educators. This site gives you an overview of the journal and will link with other sites they have. They have a mailing list and a free online discussion group. They also have a site with teaching software,

lesson plans, and the latest Internet navigations tools for the Macintosh or Windows PC.

NCTM, National Council of Teachers of Mathematics

http://www.nctm.org

This is the site for the National Council of Teachers of Mathematics and a must for all math teachers. It gives the locations and dates of regional and national math conferences as well as providing the latest news in math education . You can find a list of materials and journals.

AIMS (Activities Integrating Mathematics and Science) Education Foundation

http://www.ncrel.org/skrs/areas/issues/content/cntareas/science/sc5aims.htm

Their home page provides teachers with information abut the AIMS resource materials available. Also, a puzzle or problem of the month is posted at this Web site.

Swarthmore Math Site

http://forum.swarthmore.edu/electronic.newsletter/

This is one of the best all-round sites for mathematics. It lists super Web sites in every issue of its newsletter and has numerous links to their other sites.

NASA Online Educational Resources

http://www.gsfc.nasa.gov/nasa_online_education.html

This is the homepage for NASA Spacelink, an electronic information system for educators designed to provide current information and instructional materials related to the space program.

Source for Software With Descriptions

http://www.edsoft.com, ph.800-955-5570

The Textbook League

http://www.csulb.edu/~ttl/index.htm

This is an independent review organization for textbooks and curriculum materials and offers its bimonthly bulletin, *The Textbook Letter,* online.

Eisenhower National Clearinghouse Online

http://www.enc.org/online/ENC2670/2670.htm

This features "Promising Practices in Mathematics and Science Education 1995," highlighting 50 innovative teaching programs.

K-12 Lesson and Software Database

http://www.mste.uiuc.edu/mathed/queryform.html

This comes from the University of Illinois at Urbana-Champaign and has ready-to-use lessons that rely on the Internet to teach math.

The Maze Man

http://users.aol.com/themazeman/index.html

This site is designed to exercise mathematics skills, vocabulary, and creativity with educational puzzles.

Calculus Resources

http://archives.math.utk.edu/calculus/crol.html

Online, from Mathematics Archives, contains information on, and links to, calculus teaching and learning resources.

PBS Mathline

http://www.pbs.org/mathline/

This project is aimed at middle and elementary school students. After the teacher enrolls in PBS MATHLINE through their local public television station, the teacher-participants become members of a learning community. It is designed to implement videos and online materials to teach probability and statistics.

Circles of Light: Mathematics of Rainbows

http://www.geom.umn.edu/education/calc-init/rainbow/

This lab is designed to use a mathematical model of light passing through a water droplet to explain the development of a rainbow. It is designed for use with students in grades 9–12.

Puzzle of the Month (AIMS)

http://204.161.33.100/Puzzle/PuzzleList.html

This site includes a monthly puzzle to capture both student interest and imagination. It has some creative problems to help develop student problem-solving skills.

Centura Financial Calculators

http://www.centura.com/formulas/whatif.html

This site has online calculators to use when teaching students about automobile loans, mortgages, and investments.

Probability Central

http://www.wam.umd.edu/~pgolriz/pc/prob.html

This site includes useful data and tips related to teaching probability and statistics. You can download a free windows-based standalone version of their learning section.

E-math

http://e-math.ams.org

This is the Web site of the American Mathematical Society. This one should definitely be bookmarked on your computer. It has news, online publications, and a host of teaching information related to mathematics.

Fractals

http://math.rice.edu/

This site is great for explaining fractals to elementary and middle school students.

The History of Computers and the World Wide Web

http://chuck.ace-lab.american.edu/~ah0319a/rude/comphist.htm

This site gives a history of computers and the World Wide Web and also provides glimpses of mathematicians like Pascal, Babbage, and Turing throughout the Computer Museum Network. It has interactive exhibits and resources. This site is a good reference for a math project.

K-12 Statistics

http://www.mste.uiuc.edu/stat/stat.html

This site has lessons and data sets to use in teaching statistics. It also includes the NCTM *Statistics Standards* related to the lesson.

CyberStatistics

http://www.smoky.org/~nces/Nancy_Long/CyberStatistics.html

This site can be used to develop an interdisciplinary unit combining language arts and mathematics at the high school level. Charts and graphs are analyzed, and then techniques of technical writing are implemented to write a series of essays.

Going to a Museum

http://curry.edschool.virginia.edu/curry/

This site offers a collection of lesson plans and activities to help teachers with planning a field trip to a museum. It contains sample permission slips and lesson plans for a variety of subjects including math. It also has a list of links to online teacher resources.

Ask an Expert

http://www.askanexpert.com/askanexpert/

This site is just what it sounds like. There are experts on all kinds of subjects—from astronomy to zoology—available to answer questions from your students. This site is also a way for you to integrate the Internet into the curriculum as you teach various math topics that relate directly to a specific professional. Experts include: Ask an Architect, Ask an Astronaut, Ask a Construction Expert, Ask a Geologist, Ask a Volcanologist, and Ask Dr. Math. These have different e-mail addresses but are available on the site address above.

Math Central

http://MathCentral.uregina.ca/index.html

This is the site to go to for teachers to share resources and to find answers to difficult mathematical questions. There is Teacher Talk, a math-oriented mailing list, and a Resource Room where you will find out about Web sites and newsletters, as well as a database of math-related postings.

Lesson Plans Page

http://www.missouri.edu/~c598249/edu.html

This site has lesson plans that focus on math and science.

Chocolate Chip Cookie Math

http://sss.chipsahoy.com

You will find fun math lessons suggesting the use of chocolate chip cookies, for use in K-12 statistics.

New Math Teachers

http://www.clarityconnect.com/webpages/terri/terri.html

This site is a homepage for new math teachers offering advice, ideas, and resources.

World Wide Math Tutor

http://tqd.advanced.org/2949/index.html

If you have students that could use a little extra help, and they have access to a computer and the Internet, this might be helpful.

The Geometry Center

http://www.geom.umn.edu/welcome.html

This site is primarily for grades 7–12. It offers materials developed for high school geometry classes. The materials are for teachers and students.

Math in Oriental Rugs

http://forum.swarthmore.edu/geometry/rugs

This site is primarily for grades 7–12 geometry. It explores the symmetry of oriental rugs. Quite colorful and interesting.

Teaching Math in Elementary and Secondary Schools

http://euclid.barry.edu/~marinas/index.html

A professor at Barry University in Florida designed this Web site so that her future teachers could share their lesson plans and project ideas. Some are appropriate for math.

Algebra Online

http://www.algebra-online.com

This site has received many awards. It is a great mathematical resource for students and teachers. It has a message board, chat room, one-on-one e-mail tutoring , a mailing list, and a searchable index.

Teaching with Technology- The Well Connected Educator

http://www.gsh.org/wce/

Sponsored by the National Science Foundation, Microsoft, Compaq, and the Global Schoolhouse, this site has articles, columns, and features contributed by teachers, administrators, and parents. It is full of strategies and examples of how to use technology for teaching and learning.

Logic in Mysteries-Using Classroom Connect

http://www.mysterynet.com/

This is a lesson plan provided in *Classroom Connect* (March 1997, p. 12). It has an overview of the lesson on deriving valid arguments from given information, employing variables to symbolically represent statements, and creating a valid proof of a solved mystery case. It has questions for students to explore and refers them to a site online to help them find the answer. This would be a fun site to visit when doing proofs in geometry.

Lesson Plans Galore

http://www.kings.k12.ca.us/math/

Quite a site from our friends in California. Ignore the correlation to California SOL numbers; there is a wealth of lesson plans. So easy to use, they even have a Web-browser of their own. We searched fractions and up popped a list of fraction lesson synopses you can click on!

Geometry Problems of the Week

http://www.eduplace.com/math/brain/index.html

A fun site that has geometry problems of the week for which students can enter solutions and compete with other students. The site keeps track of who submits and announces awards for students who solve them. Grades 7 and up.

AP Statistics Resource

http://www.mindspring.com/~waus2/apstat

This site will put you in contact with resources and a discussion group about issues in AP Statistics.

Lesson and Unit Plans

http://forum.swarthmore.edu/web.units.html

Another great site that links you to unit plans. Plans are available for grades 3 through high school. You just have to read the abstract for each to find which are for you!

Virtual Math

http://www.webcom.com/~vschool

Click on the mathematics button and be introduced to "the virtual math problem," "the virtual algebra problem," "the virtual geometry problem," and "the virtual puzzle." This is for upper elementary through middle school students. You can reach archived problems also, so there is a wealth of resources here.

Site for All Teachers

http://www.education-world.com

This site is designed for teachers of all subjects and grade levels. You can get lesson plans, talk in the chatroom, and get information on the latest educational news and the hottest topics.

Texas Instruments
http://www.ti.com/calc/docs/80xthing.htm

Eightysomething! is the newsletter for users of TI graphing calculators. You can download the newsletter and any programs referenced in the issue. The following articles are referenced in the Fall 1996 issue:

"Making Connections With the World Wide Web, the TI-CBL, and TI Graphing Calculators," by Frank Wattenberg (http://www.weber.edu/math/frank-wattenberg/connected/home.htm

"Measuring the Not-So-Simple Pendulum with the CBL," by John D. Chamberlain. (He was a principal contributor to the consortium-developed CORD Applied Mathematics materials.) chamber@cord.org.

The University of Western Ontairo's Mathematics Eduation Resource Page
http://www.uwo.ca/edu/math

This site lists a variety of Web links and free software for all levels of mathematics teaching.

The Web Access Virtual Education Project
http://WAVE.uni.uiuc.edu

Located at University Laboratory High School, Urbana, IL, this site offers downloadable Mathematica notebooks for use in high school mathematics curricula.

Mathematics Hotlist
http://sln.fi.edu/tfi/hotlists/math.html

This site has an extensive list of mathematics-related Web site links.

Ask Mr. Calculus
http://www.seresc.k12.ng.us/www/apsum.html

1997 Calculus AP exam questions and solutions for AB and BC. Also, help is provided on questions submitted by the viewer.

Live From the (Hubble) Space Telescope
(http://www.quest.arc.nasa.gov/livefrom/hst.html)

Appetizers and Lessons for Math and Reason
http://www.cam.org/~aselby/lesson.html

Suggestions and problems for math as well as reflections on teaching for all disciplines. Includes logic puzzles, algebra-building skills, and a guide to stydying math.

The Guide to Math and Science Reform
http://www.hmco/college/chemistry/resourcesite/digests/chemedl/cedjan96/msg00032.htm

A database of more than 950 entries that describes projects and groups dedicated to improving K-12 mathematics and science education.

History of Mathematics
http://alephO.clarku.edu/~djoyce/mathhist/mathhist.html

For students interested in learning about ancient discoveries and mathematicians. Includes timelines, chronologies, and links to history-oriented sites.

Math Gophers
Gopher to: math.lfc.edu
look at *Mathematics Related Items*

Lists of more than 100 math-related gophers, plus archives from newsgroups, discussion lists, and bulletin boards.

Mathematics Projects
http://www.ed.hawaii.edu/

Mathematics projects that improve students' math knowledge. Current projects include algebra, geometry, and middle grades.

Newtonia
http://wwwcn.cern.ch/~mcnab/n/

Everything you always wanted to know about Sir Isaac Newton is at this site.

Online Mathematics Dictionary
http://www.mathpro.com/math/glossary/glossary.html

Definitions of basic as well as complex mathematics terms.

Pi Through the Ages
http://www-groups.dcs.st-and.ac.uk/~history/HistTopics/Pi_through_the_ages.html

This is a journey through the approximations of 3.14

21st Century Problem Solving

http://www2.hawaii.edu/suremath/home.html

Problem solving strategies for algebra, physics, and chemistry.

World of Escher

http://www.texas.net/~escher

This site discusses why many teachers use Escher's work as a tool.

WEB SITES FOUND AND CRITIQUED BY STUDENTS

The following sites were investigated and reviewed by some of our students as part of a math project at the end of the year. The students had to find either sites related to mathematics in general, or related to geometry or calculus. They visited each site, wrote a very short overview of the site, and stated whether they thought the site best served students or teachers of mathematics. The students were creative. Two students designed Internet Scavenger Hunts for the rest of the class to do as a culminating activity after completing the Web sites project presentation to the class. The sites and the student reviews are included to give you the students' perspective of the sites.

Geometry in Action

http://www.ics.uci.edu/~eppstein/geom.html

This site just talks about how geometry is used in many different things. Not very interesting to me, but may be to a geometry teacher.

Geometry of Interactive Geometry

http://www.geom.umn.edu/apps/gallery.html

This one is awesome. It has lots of things to do. I played this game. It was sort of like pinball. It was really mathematics.

Geometry Center

http://www.geom.umn.edu/

This is mainly for teachers. It has some stuff from other teachers all over the country. They have classroom activities and things like that.

Geometry Forum

http://www.forum.swarthmore.edu/

This has this thing called "Ask Dr. Math." You just simply give it a question. It wasn't very helpful, though.

Connected Geometry
http://www.edc.org/LTT/ConnGeo/

This site is run by the National Science Foundation. It's supposed to help students and teachers better understand the math concept. Explains things like proofs, area, trigonometry, coordinates, etc.

Erwin's Home Page
http://www.geocities.com/Athens/6172/main.html

This site is really just trying to sell geometry software, but it sounds really cool. It says that it teaches geometry using colors. You should really check it out. It has neat geometry link. (Very colorful!)

VRML Geometry Teacher
http://www.voicenet.com/~techno/goem.html

This has awesome pictures. Pretty good geometry info. The Mega-Graphics link has the wildest colors and pictures. Two thumbs up.

The Geometry Center Graphics Archive
http://www.geom.umn.edu/graphics/

This site has 3D pictures of many different dodecahedrons. It explains a little bit about the dodecahedrons, too.

Tucson Magnet High School Geometry Home Page
http://www.thms.k12.az.us/mathscitech/geometry/index.html

Cool geometrical shapes and pictures. Some explanations were confusing.

Math Pages
http://www.seanet.com/~ksbrown/

Extremely informative. If anyone ever has any question about anything dealing with any type of math, go HERE. You go in and it gives you a list of every type of math imaginable. Algebra, geometry, calculus, physics, theorems, History of Math—you name it! When you click on one of these, geometry for example, it gives you a list of things dealing with geometry. Points on a Sphere, the Golden Triangle, gravity sextant…etc. This is a definite must.

AP Calculus
http://www.seresc.k12.nh.us/www/ apsum.html

A great site for teachers and students preparing to take the AP Calculus exam. It offers a list of summer workshops for teachers, a current syllabus and a summary of the changes in the 1998 course description, a list of approved graphing calculators, and calculus resources.

The Remote Learning Project

http://www.math.ohio-state.edu/~davis/coursedes.html

Currently this site has two good projects. Internet for Teachers is designed to help teachers integrate the internet into their teaching. The next project, The Grape Vine, is a network for students and teachers intended for the sharing of projects and problems. The Grape Vine links to our next site and deserves an entry of its own.

The Grape Vine

http://socrates.math.ohio-state.edu/

This is a very cool site with lots of links to other sites. It offers a news group, a way to turn in your homework via the Internet, and even monitor your grades for this online course. This is the best site I have seen for math.

Calculus and Mathematica

http://www.-cm.math.uiuc.edu/index.html

This site offers a new and exciting way to learn calculus, differential and matrix theory. This course emphasizes experimentation and real-time visualization to learn fundamental calculus concepts and ideas. The site also has a very nifty background graphic.

Graphics for the Calculus Classroom

http://www.math.psu.edu/dna/graphics.html

This site has a series of graphic demonstrations, developed to supplement a first year calculus class. It has great graphics.

Fairfield High School AP Calculus Home Page

http://www.geocities.com/CapeCanaveral/Lab/2619/

This site looks like a project from this AP Calculus Class. It begins with a funny definition that is worth reading and offers a bibliography, problems, and a funny quiz called "Are You a Math Geek?" The site doesn't contain much mathematically, but the section of quotes it contains is an absolute must for any beginning calculus student. This site is definitely a student site.

Calculus@internet

http://www.calculus.net/

A title page with links to many calculus area-labs, homework, VRML, texts, projects, history, differential, integral, vector, ode, linear, and precalculus. This site also offers a bookstore as a way to buy many of the referenced works. This site contains every thing under the sun and then some. This site definitely leans towards being a teacher site.

Math Mania

http://csr.uvic.ca/~mmania/

This site has information concerning graphs, knots, networks, and finite state machines. Each section offers an array of options—puzzles, tutorials, activities, applications, research, and more. This site is most definitely a student site.

Interactive Mathematics Online

http://tqd.advanced.org/2647/index.html

This site has algebra, trig, geometry, Chaos, and lots of Cool Java stuff. The Java stuff isn't really math related but is still really cool. Each section gives you a brief introduction, a reason to study, and then a tour including examples.

Virtual Polyhedra

http://www.li.net/~george/virtual-polyhedra/vp.html

A great geometry site! It has 2D pictures and descriptions of Platonic Solids, Kepler-Poinsot polyhedra, prisms and antiprisms, Archemedian duals, and more.

The Math and Physics Help Home Page

http://www2.ncsu.edu/unity/lockers/users/f/felder/public/kenny/home.html

Good in-depth help sheets for math and physics. It has some funny intro pieces including a song "Failing my Calculus" sung to the tune of "Closer to Fine" by the Indigo Girls. The author uses humor and funny stories to get your attention. Definitely a student site but one that cool teachers will like.

SOS Mathematics

http://www.math.utep.edu/sosmath/

A great site. This site has review material for algebra, trig, calculus, differential equations, complex variables, and matrices. The algebra section has a number of topics complete with rules, examples and problems for you to work. This site also has a nifty background.

AMATH—Prealgebra, Basic Math Skills
http://www.amath.com/index.html

This site is a tutorial for adults who have completed K-8 math but can't do the work. It covers arithmetic, measurements, algebra, geometry, and data analysis. This site is a good site for students who have problems showing their work but, of course, we don't know any one like that!

UMass/Amherst GANG Home Page
http://www.gang.umass.edu/

This is an interdisciplinary differential geometry research team. The opening isn't great but once you get into the rest of the page it is amazing. It requires you to download before you can fully view the figures. This site is most definitely a teacher site or a very advanced student site.

Fractal Geometry
http://www.idg.fhg.de/~kumpf/fraggels/v_peano.gif

This is a small site dealing with fractal geometry. The fractals they show are very nifty. I couldn't really say much about the quality of the info because we haven't done fractal geometry. This is a good site for students and teachers.

The Integrator
http://www.integrals.com/

This site allows you to enter any differentiation and it will solve it for you. This site would be very helpful on math problems that students couldn't figure out what the answer was. This site is more of a student site than a teacher site.

Finite Math and Calculus
http://www.hofstra.edu/~matscw/realworld.html

This site offers interactive tutorials, quizzes, review exercises, and tests for logic, trig, and probability and statistics. The quizzes correspond to a set of textbooks. This site would be helpful for both teachers and students.

Appetizers and Lessons for Math and Reasoning

http://www.cam.org/~aselby/lesson.html

This is a pretty standard website—gray background, highlighted text, boxes, etc. I found it rather drab but some people like sites like that. It offers appetizers for basic math reasoning and writing. This site is intended for students but I think teachers would have more luck using it because students would get bored.

The Equality Sisters

http://www-leland.stanford.edu/~meehan/xyz/equality.html

This site tries for humor and has very few math problems. Any competent student can skip this site but for those students struggling with algebra, I would recommend it. Definitely not a teacher site, although a few might find it humorous.

K-12 Math Education

http://MathCentral.Uregina.ca/

This is more like a general chat room about math, but it does have a database in the resource room. This site can be used by either students or teachers if they like chatting or need to look something up.

INTERNET RESOURCES ON BLOCK SCHEDULING

http://www.kusd.edu/district/whatshappening/initiatives/blocksched.html
http://www.classroom.net/classweb/WASSON/myhome.html
http://www3.classroom.net/classweb/WASSON/resources.html
http://www.stmary.k12.la.us/block.html
http://www.drpa.org:81/~lion/block.html

APPENDIX D

REFERENCES

Aichele, Douglass B. (Ed.). (1994). *Professional Development for Teachers of Mathematics.* Reston, VA: National Council of Teachers of Mathematics.

American Mathematical Association of Two-Year Colleges (1995). *Crossroads in Mathematics: Standards for Introductory College Mathematics Before Calculus.* Memphis, TN: The Association.

Anderson, Richard, Brozynski, Bruce, & Lett, David. (1996). Scheduling with purpose: Key people and key objectives. *High School Magazine,* March/April, 24–26.

Artzt, Alice F., & Newman, Claire M. (1990). Cooperative learning. *Mathematics Teacher,* 83(9), 448–52.

Artzt, Alice F., & Newman, Claire M. (1990). *How to Use Cooperative Learning in the Mathematics Classroom.* Reston, VA: National Council of Teachers of Mathematics.

Assessment Standards Working Groups of the NCTM. (1995). *Assessment Standards for School Mathematics.* Reston, VA: National Council of Teachers of Mathematics.

Austin, Joe Dan (Ed.). (1991). *Applications of Secondary School Mathematics Readings from the Mathematics Teacher.* Reston, VA: National Council of Teachers of Mathematics.

Burrill, Gail et al. (1992). *Data Analysis and Statistics across the Curriculum: Addenda Series, Grades 9–12.* Reston, VA: National Council of Teachers of Mathematics.

Canady, Robert Lynn, & Rettig, Michael D. (1995). The power of innovative scheduling. *Educational Leadership,* 53(3), 4–10.

Canady, Robert Lynn, & Rettig, Michael D. (1995). *Block Scheduling: A Catalyst for Change in High Schools.* Princeton, NJ: Eye On Education.

Canady, Robert Lynn, & Rettig, Michael D. (1996). *Teaching in the Block: Strategies for Engaging Active Learners*. Princeton, NJ: Eye On Education.

Carroll, J.M. (1994). *The Copernican Plan Evaluated: The Evolution of a Revolution.* Topsfield, MA: Copernican Associates.

Carl, Iris M. (1989). Essential mathematics for the twenty-first century: The position of the National Council of Supervisors of Mathematics. *Mathematics Teacher,* 82(5), 388–91.

Cawelti, G. (1997). *Effects of High School Restructuring: Ten Schools at Work.* Arlington, VA: Educational Research Service.

Charles, Randall, Lester, Frank, & O'Daffer, Phares. (1992). *How to Evaluate Progress in Problem Solving.* Reston, VA: National Council of Teachers of Mathematics.

College Board (1996a, September 19 (revised March 10, 1997)). AP and January examinations. (Press release). Princeton, NJ: (Author).

College Board (1996b, December 12). Performance of AP students who are block scheduled. Princeton, NJ: (Author).

Cooney, Thomas J., & Hirsch, Christian R. (Eds.) (1990). *Teaching and Learning Mathematics in the 1990s (1990 Yearbook).* Reston, VA: National Council of Teachers of Mathematics.

Corbitt, Mary Kay. (Ed.). (1985). The impact of computing technology on school mathematics: Report of an NCTM conference. *Mathematics Teacher,* 78(4), 243–50.

Coxford, Arthur E. (Ed.). (1988). *The Ideas of Algebra, K-12 (1988 Yearbook).* Reston, VA: National Council of Teachers of Mathematics.

Coxford Jr., Arthur F., with Burks, Linda, Giamati, Claudia, & Jonik, Joyce. (1991). *Geometry from Multiple Perspectives. Curriculum and Evaluation Standards for School Mathematics: Addenda Series, Grades 9–12.* Reston, VA: National Council of Teachers of Mathematics.

Cuevas, Gilbert, & Driscoll, Mark. (Eds.). (1993) *Reaching All Students with Mathematics.* Reston, VA: National Council of Teachers of Mathematics.

Cunningham Jr., R. Daniel, & Nogle, Sue Ann. (1996). Implementing semesterized schedule: Six key elements. *High School Magazine,* March/April, 28–32.

Davidson, Neil. (1990). *Cooperative Learning in Mathematics, A Handbook for Teachers.* Menlo Heights, CA: Addison-Wesley.

Day, Roger. (1996). Classroom technology: Tool for, or focus of, learning? *Mathematics Teacher*, 89(2), 134–37.

Demana, Franklin, & Waits, Bert K. (1990). The role of technology in teaching mathematics. *Mathematics Teacher*, 83(1), 27–31.

Edwards Jr., C.M. (1995). The 4x4 plan. *Educational Leadership*, 53 (3) 16–19.

Elliott, Portia C. (Ed.). (1996). *Communication in Mathematics, K-12 and Beyond (1996 Yearbook)*. Reston, VA: National Council of Teachers of Mathematics.

Fey, James T., & Hirsh, Christian R. (1992). *Calculators in Mathematics Education, (1992 Yearbook)*. Reston, VA: National Council of Teachers of Mathematics.

Froelich, Gary W., with Bartkovich, Kevin G., and Foerster, Paul A. (Eds.). (1991). *Curriculum and Evaluation Standards for School Mathematics: Addenda Series Grades 9–12*. Reston, VA: National Council of Teachers of Mathematics.

Froelich, Gary W., with Bartkovich, Kevin G., & Foerster, Paul A. (Eds.) (1995). *Connecting Mathematics across the Curriculum (1995 Yearbook)*. Reston, VA: National Council of Teachers of Mathematics.

Garfunkel, Solomon (Project Director). (1991). *For All Practical Purposes, Introduction to Contemporary Mathematics*. 2nd ed. COMAP, Inc. New York: W.H. Freeman and Company.

Geddes, Dorothy et al. (1994). *Measurement in the Middle Grades: Addenda Series, Grades 5–8*. Reston, VA: The National Council of Teachers of Mathematics.

Governor Thomas Johnson High School. (1995). *Three-year Summary Report 1992–95*. Frederick, MD: Author, (301) 694-1412.

Gunter, Mary Alice, Estes, Thomas H., & Schwab, Jan Hasbrouck. (1990). *Instruction: A Models Approach*. Boston, MA: Allyn and Bacon.

Hackman, D.G. (1995). Ten guidelines for implementing block schedule. *Educational Leadership*, 53, (3), 24–27.

Heid, M. Kathleen (1988). Calculators on tests: One giant step for mathematics education. *Mathematics Teacher*, 81(9), 710–13.

Heid, M. Kathleen, Choate, Jonathan, Sheets, Charlene, & Zbiek, Mary Ann. (Eds.) (1995). *Algebra in a Technological World: Addenda Series, Grades 9–12*. Reston, VA: National Council of Teachers of Mathematics.

Hiatt, Arthur A. (1987). Discovering mathematics. *Mathematics Teacher*, 80(9), 476–77.

Hirsch, Christian R., & Laing, Robert A. (1993). *Activities for Active Learning and Teaching: Selections from the Mathematics Teacher.* Reston, VA: National Council of Teachers of Mathematics.

Hottenstein, D. and Malatesta, C. (1993). Putting a school in gear with intensive scheduling. *High School Magazine,* 1(2), 28–29.

International Reading Association. (1976). *Teaching and Reading Mathematics.* Newark, DE: The Association.

Johnson, Bil. (1996). *The Performance Assessment Handbook, Volume 1: Portfolios & Socratic Seminars.* Princeton, NJ: Eye on Education.

Johnson, Bil. (1996). *The Performance Assessment Handbook, Volume 2: Performances & Exhibitions.* Princeton, NJ: Eye on Education.

Kagan, Spencer. (1994). *Cooperative Learning.* San Juan Capistrano, CA: Resources for Teachers, Inc.

Keese, Earl E. (1975). "The Pit and the Pendulum": Source for a creative activity. *Mathematics Teacher,* 68(7), 602–04.

Kenney, Margaret J. (Ed.). (1991). *Discrete Mathematics Across the Curriculum, K-12.* Reston, VA: National Council of Teachers of Mathematics.

Kern, Cherlyn. (1997). Descriptive-paragraph mini project. *Mathematics Teacher,* 90(5), 362–63.

Kobrin, David. (1992). *In There With the Kids, Teaching in Today's Classrooms.* Boston: Houghton Mifflin Company.

Kramer, S.L. (1996). Block scheduling and high school mathematics instruction. *Mathematics Teacher,* 89(9), 758–67.

Language & Reading Mathematics Group. (1984). *Children Reading Mathematics.* London, England: The Group.

Larson, Ronald E., Kanold, Timothy D., & Stiff, Lee. (1993). *Algebra I.* Lexington, MA: D.C. Heath and Company.

Larson, Ronald E., Kanold, Timothy D., & Stiff, Lee. (1993). *Algebra II.* Lexington, MA: D.C. Heath and Company.

Lindquist, Mary Montgomery. (Ed.). (1987). *The Learning and Teaching Geometry, K-12 (1987 Yearbook).* Reston, VA: National Council of Teachers of Mathematics.

Matthews, Peter. (Ed.). (1996). *The Guinness Book of Records 1996.* New York, NY: Bantam Books.

Matras, Mary Ann. (1991). Technology in the classroom: beginnings and endings. *Mathematics Teacher,* 84(2), 86–87.

McDonald, Janet L. (1988). Integrating spreadsheets into the mathematics classroom. *Mathematics Teacher,* 81(11), 615–19.

Meiring, Steven P., Rubenstein, Rheta N., Schultz, James E., de Lange, Jan, & Chambers, Donald L. (1992). *A Core Curriculum—Making Mathematics Count for Everyone: Addenda Series, Grades 9–12.* Reston, VA: National Council of Teachers of Mathematics.

Mitchell, Charles E. (1989). Henry Wadsworth Longfellow, poet extraordinaire. *Mathematics Teacher,* 82(5), 378–79.

Nahrgang, Cynthia L., & Petersen, Bruce T. (1986). Using writing to learn mathematics. *Mathematics Teacher,* 79(6), 461–65.

National Center for Improving Science Education. (1996). *Bold Venture, Volume 3, Case Studies of U.S. Innovations in Mathematics Education.* Boston, MA: Kluwer Academic Publishers.

National Council of Teachers of Mathematics. (1989). *Curriculum and Evaluation Standards for School Mathematics.* Reston, VA: National Council of Teachers of Mathematics.

Neil, Marilyn S. (1996). *Mathematics The Write Way: Activities for Every Elementary Classroom.* Princeton, NJ: Eye On Education.

O'Neil, J. (1995). Finding time. *Educational Leadership,* 53(3), 11–15.

Owen, Lisa B., & Lamb, Charles E. (1996). *Bringing the NCTM Standards to Life.* Princeton, NJ: Eye On Education.

Phillips, Elizabeth, with Gardella, Theodore, Kelly, Constance, & Stewart, Jacqueline. (1991). *Patterns and Functions, Curriculum and Evaluation Standards for School Mathematics: Addenda Series, Grades 5–8.* Reston, VA: National Council of Teachers of Mathematics.

Pugalee, David K. (1997). Connecting writing to the mathematics curriculum. *Mathematics Teacher,* 90(4), 308–10.

Reys, Barbara J. (1991). *Developing Number Sense in the Middle Grades: Addenda Series, 5–8.* Reston, VA: National Council of Teachers of Mathematics.

Robbins, Christine Black, & Geiger, Phil. (1996). Ways to schedule the school day. *High School Magazine,* March/April, 21–23.

Salvaterra, M., & Adams, D. (1995). Departing from tradition: two schools' stories. *Educational Leadership*, 53(3) 32–36.

Schoen, Harold L. (1989). Beginning to implement the Standards in Grades 7–12. *Mathematics Teacher*, 82(9), 427–30.

Sessoms, J.C. (1995). *Teachers perceptions of three models of high school block scheduling*. Unpublished doctoral dissertation, University of Virginia, Charlottesville.

Serra, Michael. (1993). *Discovering Geometry, An Inductive Approach*. Berkeley, CA: Key Curriculum Press.

Serra, Michael. (1993). *Patty Paper Geometry*. Berkeley, CA: Key Curriculum Press.

Stenmark, Jean Kerr. (Ed.). (1991). *Mathematics Assessment: Myths, Models, Good Questions, and Practical Suggestions*. Reston, VA: National Council of Teachers of Mathematics.

Stumpf, T. (1995). A Colorado school's un-rocky road to trimesters. *Educational Leadership*, 53(3), 20–23.

Troyka, Lynn Quitman. (1990). *Simon & Schuster Handbook for Writers*. Englewood Cliffs, NJ: Prentice Hall.

Webb, Norman L. (1993). *Assessment in the Mathematics Classroom (1993 Yearbook)*. Reston, VA: National Council of Mathematics Teachers.

Wilkinson, Jack D. (1984). The computer: a tool for instruction? *Mathematics Teacher*, 77(6), 404, 405, 490.

Wilson, Patricia S. (1993). *Research Ideas for the Classroom, High School Mathematics*. New York: Macmillan Publishing Company.

Working Group chaired by Ball, Deborah, Cooney, Thomas, & Friel, Susan. (1991). *Professional Standards for Teaching Mathematics*. Reston, VA: National Council of Teachers of Mathematics.

Zawojewski, Judith S. et al. (1991). *Dealing with Data and Chance: Addenda Series, Grades 5–8*. Reston, VA: National Council of Teachers of Mathematics.

For Product Safety Concerns and Information please contact our EU representative GPSR@taylorandfrancis.com Taylor & Francis Verlag GmbH, Kaufingerstraße 24, 80331 München, Germany

Batch number: 10733528

Printed by Printforce, the Netherlands